Meet the Industry Candidates Poster Session 2019

Held at the 2019 AIChE Spring Meeting and
15th Global Congress on Process Safety

New Orleans, Louisiana, USA
31 March - 4 April 2019

ISBN: 978-1-5108-8680-3

Printed from e-media with permission by:

Curran Associates, Inc.
57 Morehouse Lane
Red Hook, NY 12571

Printed by Curran Associates, Inc. (2019)

For permission requests, please contact AIChE
at the address below.

AIChE
120 Wall Street, FL 23
New York, NY 10005-4020

Phone: (800) 242-4363
Fax: (203) 775-5177

www.aiche.org

Additional copies of this publication are available from:

Curran Associates, Inc.
57 Morehouse Lane
Red Hook, NY 12571 USA
Phone: 845-758-0400
Fax: 845-758-2633
Email: curran@proceedings.com
Web: www.proceedings.com

TABLE OF CONTENTS

Crystallization Kinetic Measurement and Parameter Estimation Utilizing Population Balance Model in a Dynamic/Oscillatory Baffle Crystallizer

Claire Y. Liu†, Alexia†, Alastair Barton‡, Paul Firth‡, Prof. Zoltan K. Nagy†
†School of Chemical Engineering, Purdue University ‡Alconbury Weston Ltd, Stoke-on-Trent, UK

Continuous Manufacturing of Pharmaceuticals

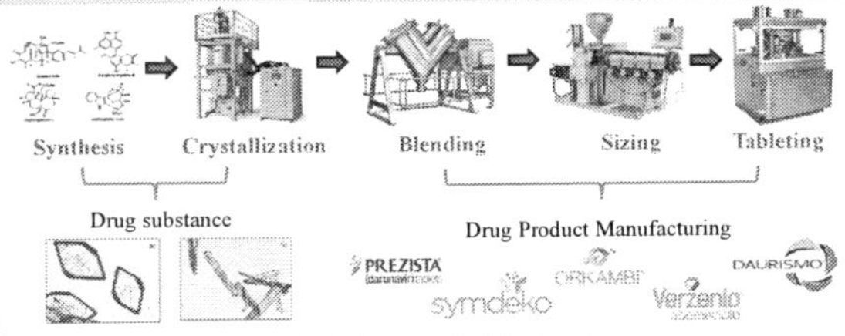

Figure 1. Typical Pharmaceutical Manufacturing

- Crystallization plays the transitional role followed by downstream drug product manufacturing.
- Effective crystal engineering lowers down-stream processing difficulty and even eliminate some unit operations → economical
- Mechanistic understanding is essential to understanding the process → QbD
- Continuous manufacturing, identified as an emerging technology, relies on kinematic understanding of the process.

Simone et al. (2014) *Chem. Eng. Technol.*, *37*(8), 1305–1313

Crystallization

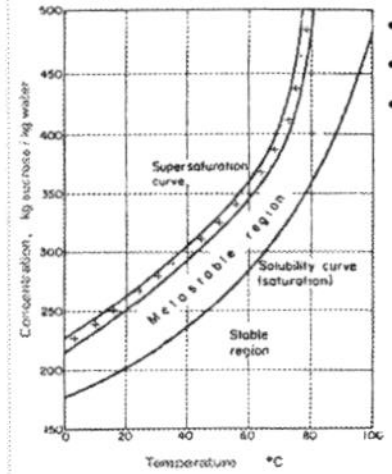

Figure 2. Phase diagram

- Nucleation: birth of crystals
- Growth: enlargement of crystals
- Supersaturation generation, seeding strategy and mixing dynamics affects crystallization kinetics

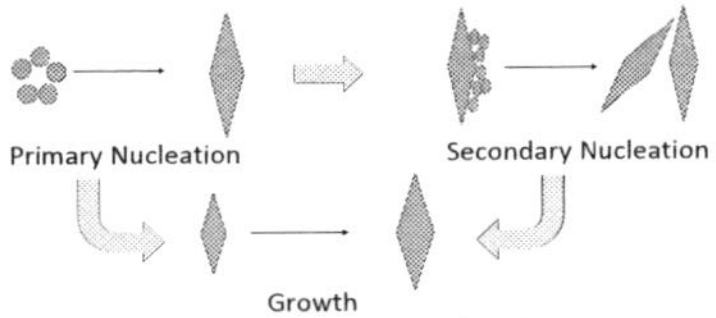

Figure 3. Crystallization kinetics schematic

Population Balance Model (PBM)

- Population balance tracks the number particles of different sizes
- Population distribution function $f(t,L)$ → fdL = # of particles of size L to $L+dL$
- Population balance

$$\frac{\partial f}{\partial t} + \frac{\partial (Gf)}{\partial L} = B\delta(L - L_0)$$

Mass balance

$$\frac{dc}{dt} = -3k_v\rho_c \int_{L_0}^{L_{max}} GL^2 f dL$$

- Batch operation
- Kinetic equations are semi-empirical

Primary nucleation: $J = k_j \exp\left(\frac{j}{T^3 \ln^2(C/C_{sat})}\right)$

$J = 0$ if seeded, common technique

Secondary nucleation: $B = k_b \left(\frac{C}{C_{sat}}\right)^b \exp\left(\frac{-E_b}{RT}\right) M^m$

Growth: $G = k_g \left(\frac{C}{C_{sat}}\right)^g \exp\left(\frac{-E_g}{RT}\right) (1 + \alpha_g L)^{L_g}$

- Initial condition
  - $F(t=0,L)$ Seed population distribution
  - $C(t=0)$ Initial solution concentration

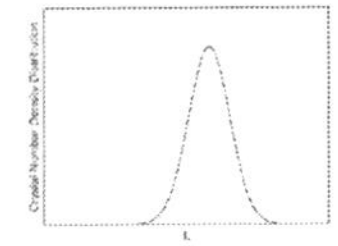

Figure 4. Population distribution

L: size$[=]\mu m$

B: nucleation rate$[=]\frac{\#}{min \cdot mL}$

G: growth rate$[=]\frac{\mu m}{min}$

Oscillatory/Dynamic Baffle Crystallizer

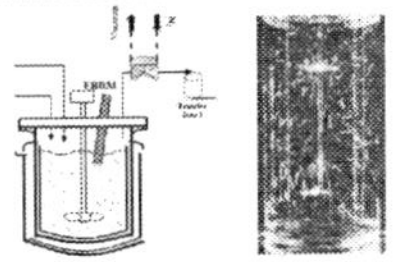

- Dynamic Baffle Crystallizer Vs Stirred Tank
  - ✓ Both can be used as Batch or Mixed-Suspension-Mixed-Product-Removal(MSMPR) continuous platforms

- Dynamic Baffle Crystallizer (DBC)
  - ✓ Improved heat and mass transfer.
  - ✓ Lessened sheer: gentler mixing
  - ✓ Improved RTD uniformity
  - ○ Noisier PAT measurement
- Stirred Tank Crystallizer
  - ✓ Utilize existing equipment
  - ✓ Easy incorporation of PATs
  - ○ Poor local mixing

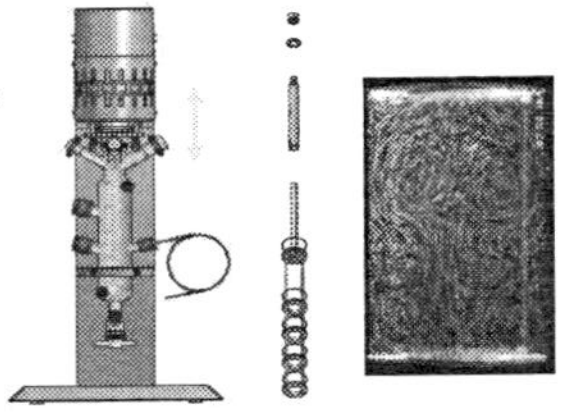

Figure 5. Stirred tank Vs. DBC

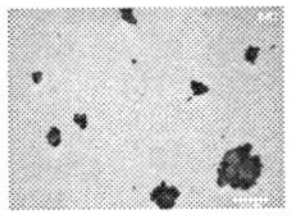

STC: agglomerated
- Solvent incorporation
- Bioavailability

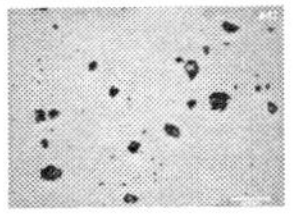

DBC: more uniform

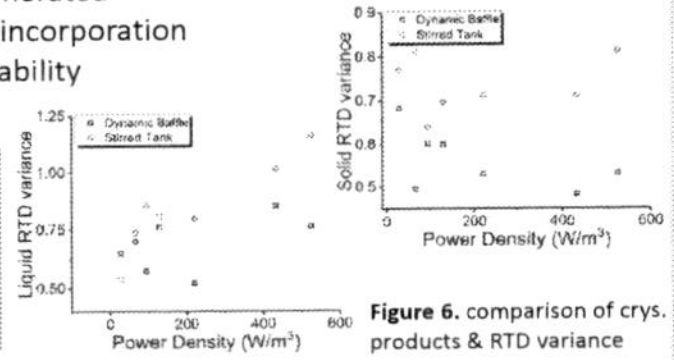

Figure 6. comparison of crys. products & RTD variance

Experimental Design

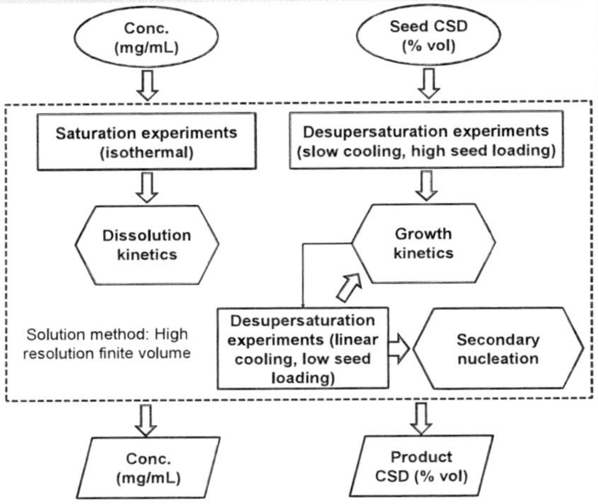

Figure 7. Systematic experimental design to decouple nucleation and growth

Parameter Estimation Results

- Growth: slow cooling, high seed loading
  - Concentration decrease: measured by UPLC
  - Particle count, measured by FBRM, does not increase → likely no nucleation
  - Square weighted mean chord length (SWMCL), measured be FBRM, increase
  - Can assume growth dominates, $B = 0$.

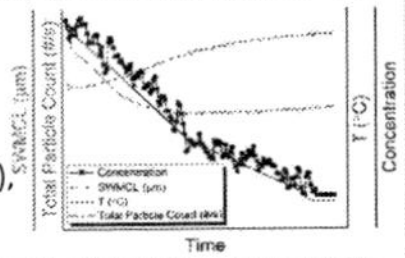

Figure 8. Growth Exp Operating Profile

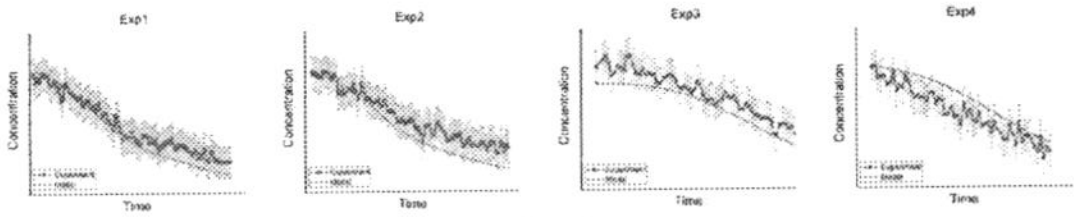

Figure 9. Growth experiments concentration fitted results

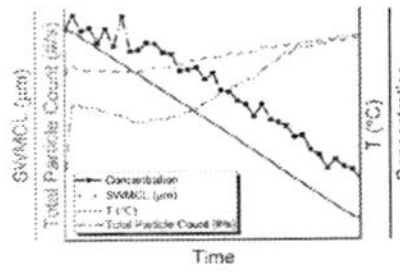

Figure 10. Growth Exp Operating Profile

- Secondary nucleation: fast cooling, low seed loading
  - Particle count increase → nucleation
  - SWCMCL decrease → number of small particles increased due to nucleation
  - Growth is always probable when supersaturation is present
  - Assume afore-estimated growth kinetics

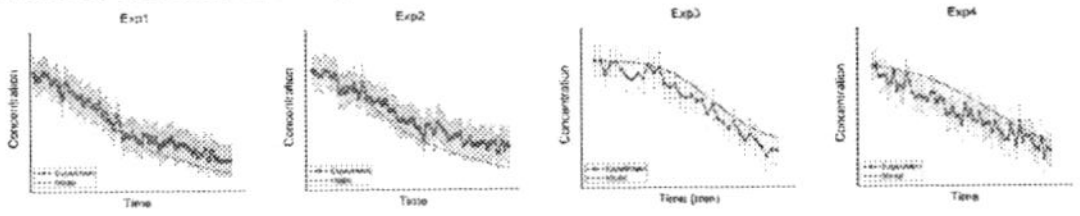

Figure 11. Secondary nucleation experiments concentration fitted results

Future Work

- Uncertainty and sensitivity analysis
- Model application in continuous crystallization
  - Start-up optimization: shorten start up time, improves yield
  - High level control: state of control instead of purely steady state

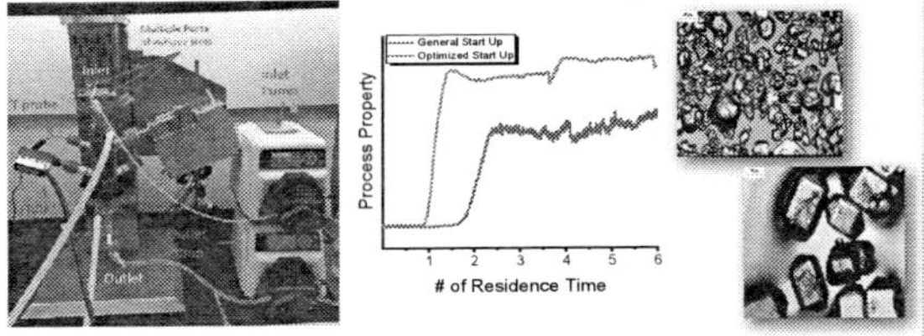

Figure 9. (a)Continuous crystallization experimental set up (b) start up optimization

Acknowledgement

Alconbury Weston Ltd
Continuous Processing Solutions
Designed around you

NiTech Solutions

P2SAC

Su et al. (2015) *Chem. Eng. Process.*, *89*, 41–53.

Claire Yiqing Liu

2101 Cumberland Ave. Apt# 2201, West Lafayette, IN 47906
(608) 213-5034
ClaireLiuPurdue@gmail.com

Career Summary

A results-oriented engineer specialized in separation process development. Looking for an opportunity to use computational skill set to execute high quality, high impact projects while exceeding project goals and profit expectation under strict deadlines and budget.

Education

Purdue University, Graduate School, West Lafayette, IN Aug. 2019
Ph.D., Davidson School of Chemical Engineering
University of Wisconsin-Madison, Madison, WI Jul. 2015
Bachelor of Science in Chemical Engineering and Mathematics, GPA:3.4

Research Experiences

Graduate Research Assistant, Dr. Zoltan K. Nagy's Lab, Purdue University **Aug. 2015 – present**
Crystallization process development of drug substance in an oscillatory system

- Developed reliable continuous crystallization processes in the unique oscillatory crystallizer and applied chemical engineering principles to compare with a traditional agitation system.
- Developed or modified offline characterizations methods to quantify critical quality attributes (CQA).
- Coupled with continuous filtration for post-synthesis drug substance continuous manufacturing.
- Calibrated and applied analytical instruments such as the Raman, FBRM, UV, and UPLC for monitoring.
- Constructed a population balance model for startup studies of continuous crystallization.

Process system modeling in micro-scale devices

- Measured crystallization kinetics for micro-scale process system modeling.

ORISE Fellow, U.S. Food and Drug Administration, Silver Spring, MD **Mar. – Aug. 2018**
Population balance model development for crystallization of carbamazepine

- Designed and performed efficient systems of experiment to extract crystallization kinetic information.
- Estimated kinetic parameters from experimental data to construct a crystallization model.
- Verified and validated the model by a risk-based approached.
- Identified critical design parameters by combined local/global sensitivity analysis (Morris Method).
- Aided design space determination by simulating the critical parameter space (model aided QbD approach).
- Transferred model by building flow sheet and custom kinetics in gPROMS for knowledge centralization.

Risk analysis of plug-flow oscillatory baffled crystallizer for pharmaceutical applications

- Constructed fish-bone risk diagrams to summarize the risk considerations of the tubular flow crystallizer.
- Evaluated continuous crystallization performance and associated risks with the aid of DOE and ANOVA.

Undergraduate Research Assistant, University of Wisconsin-Madison **Feb. 2013 – May 2015**
Bio-reaction network simulation at Dr. Hiroshi Maeda's Lab at Botany department

- Simulated lignin production network in tree barks using MATLAB to predict and improve lignin generation.

Machine Learning with Prof. Sangtae Kim and Dr. Pedro Romero

- Used machine learning algorithms to predict protein inhibitor selectivity from structural data.

Publications

Liu, Y. C.; Dunn, D.; Lipari, M.; Barton, A.; Firth, P.; Speed, J.; Wood, D.; Nagy, Z. K. A Comparative Study of Continuous Operation between a Dynamic Baffle Crystallizer and a Stirred Tank Crystallizer. *Chem. Eng. J.* **2019**, *367*, 278–294.

Acevedo, D.; Yang, X.; **Liu, Y. C.**; O'Connor, T. F.; Koswara, A.; Nagy, Z. K.; Cruz, C. N. Encrustation in continuous Pharmaceutical Crystallization Process- A Mini Review. *Org. Process Res. Dev.* 2019 (In review)

Liu, Y. C.; Acevedo, D.; Yang, X.; Naimi, S.; Wu, W.; Pavurala, N.; Nagy, Z. K.; O'Connor, T. F.; Cruz, C. N. Population Balance Model Development Verification and Validation of Crystallization of CBZ. (in preparation)
Liu, Y. C.; Mazumder, S.; Acevedo, D.; Yang, X.; Naimi, S.; Pavurala, N.; Nagy, Z. K.; O'Connor, T. F.; Cruz, C. N. Risk Considerations for Continuous Cooling Crystallization of Carbamazepine in a Plug Flow Oscillatory Baffled Crystallizer (In Preparation)
Yang, X.; Acevedo, D.; **Liu, Y. C.**; O'Connor, T. F.; Cruz, C. N. Continuous Crystallization: Quality Considerations. **Book chapter** (Submitted for clearance in DPQR at the FDA to submit to book author)
Liu, Y. C.; Nagy, Z.K.; Continuous Crystallization: Equipment and Operation. **Book chapter** (Submitted to editor)

Conference presentations

Y. C. Liu, P. Firth, A. Barton, J. Speed, D. Wood, Z.K. Nagy, A Comparative Study of Residence Time Distribution and Cooling Crystallization in a Continuous Dynamic/Oscillatory Baffle Crystallizer Versus a Stirred Tank, *2018 AIChE Annual Meeting*, Pittsburgh, PA USA, October 28-November 2, 2018
Y. C. Liu, Z. K. Nagy, G. Reklaitis Introduction to Small Molecule Oral Dosage Form Drug Manufacturing, *FDA Fellows Association 2018 Summer Intern Seminar Series*, June 29, 2018
Y. C. Liu, P. Firth, A. Barton, J. Speed, D. Wood, Z.K. Nagy, Comparative Studies of Continuous Crystallization in an Oscillatory Baffled Crystallizer (OBC) and a Mixed-Suspension-Mixed-Product-Removal Stirred Tank Crystallizer, *2017 AIChE Annual Meeting*, Minneapolis, MN, USA, October 29-November 3, 2017

Leadership

Safety Officer, School of Chemical Engineering, Purdue University **Sept. 2017 –**
- Led the research group to higher safety standards in compliance with departmental, local, state and federal regulations and policies while promoting continuous improvements.
- Provided safety training for graduate and undergraduate researchers as well as visiting scholars.
- Continuing maintenance and reinforcement of inherently safe practices and housekeeping.

Graduate Mentor, Davidson School of Engineering, Purdue University **May 2016 -**
- Mentored five undergraduate research assistants to form and integrate their independent experiment-based research projects with group research objectives while guiding them through the development of research plans as well as personalized SOPs for day-to-day work.
- Mentored two first year graduate students and provided guidance through their first year at Purdue.

Purchasing Officer, School of Chemical Engineering, Purdue University **May 2016 –**
- Communicated with departmental procurement office to ensure the group's purchasing needs.
- Ordered general items to aid the day-to-day operations in the lab while overseeing all purchasing activities.

Ambassador, Graduate Women Group, Purdue University **Aug. 2016 -**
- Initiated the outreach program at Almost Home Humane Society for monthly volunteering.
- Planned and organized monthly gatherings and semester-end get-togethers.

Advisor, DPQR, CDER, U.S. Food and Drug Administration **Mar. 2018 – Aug. 2018**
- Led a team of two undergraduate research fellows to perform experimental subprojects, guiding them through research plan development, regular team updates, and theoretical concept improvement.
- Developed protocols and SOPs for their responsible laboratory equipment.

Co-curricular chair, ChemE Graduate Student Organization, Purdue University **May 2016 – May 2017**
- Planned and organized professional development activities as well as helping to organize the symposium.

Skills

Language:	Native in Mandarin Chinese, Fluent in English (>10yrs), Learning Spanish
Technical:	MATLAB, gPROMS, JMP, LabVIEW, SAS, R, FBRM, Raman, PVM, Malvern, Machine learning
Interest:	Volunteering, dancing, yoga, real estate, travel

553213 Accumulative Roll Bonding (ARB)-Processed High-Manganese Twinning Induced Plasticity (TWIP) Steel with Extraordinary Strength and Reasonable Ductility

Accumulative roll bonding (ARB) process at room temperature was used to fabricate a nanostructured twinning-induced plasticity (TWIP) steel. The effect of the number of ARB cycles from 1 to 3 on the microstructure and mechanical properties of Fe-31Mn-3Al-3Si (wt%) TWIP steel was investigated. The results showed that the microstructure of the 1-cycle ARB-processed sample mainly contained the primary and secondary mechanical twins. However, in the microstructure of 2-cycle ARB-processed sample, new configurations of small nanotwins which they are known as the hierarchical nanotwinned (HNT) structures were observed between the secondary mechanical twins. In the microstructure of 3-cycle ARB-processed sample, new subdivided grains with a mean size of about 100 nm were produced. Additionally, the 3-cycle ARB-processed TWIP steel sample exhibited extraordinary strength (about 1.2 GPa) and reasonable ductility (about 15%). The mechanical behavior of ARB-processed TWIP steel was interpreted as the evidence for microstructural features such as the mechanical twins (especially, the HNT structures) and nano/ultrafine grains.

553316 Transforming Biomass-Derived Oxygenates Using Carbide Catalysts

As a chemical engineer, I always aspire to apply my problem-solving skills to address challenges in our lives. In this poster, all the projects that I am actively involved in during my doctoral study are integrated to demonstrate the thought processes as well as the skills that I developed. I believe that the fundamental insights that I gained through these studies should be important for the rational design of efficient catalysts for biomass conversion and industrial catalytic processes in general. The methodology and problem-solving skills used in the catalyst discovery processes can also be applied to relevant fields of chemical engineering. For presentation purposes, this poster presentation will be divided into two parts: the discussion of research results and the skills learned behind the results.

First of all, the motivation and the results of my research are discussed. My research is focused on the upgrading of biomass, which has been considered as a promising approach to reduce our current reliance on fossil fuels. Biomass is renewable and can be converted into value-added fuels and chemicals, and therefore help to address the current energy and environmental issues. Among the various technologies developed for upgrading biomass, the low temperature catalytic conversion of lignocellulosic biomass and the biodiesel production are particularly promising. During the low temperature catalytic conversion, lignocellulosic biomass can be converted into a platform chemical, furfural, which is further upgraded into fuel-additives or other valuable chemicals.[1] Molybdenum carbide (Mo_2C) catalysts have been shown to be highly active and selective in the conversion of furfural but suffers from quick deactivation. In the biodiesel production process, glycerol is generated as a major by-product. Mo_2C can potentially be applied for the upgrading of glycerol via the hydrodeoxygenation reaction, but the issue of unselective C-O bond scission should be addressed. In this work, we have identified a class of metal-modified molybdenum carbide catalysts to solve the problems in these two processes. With the decoration of cobalt, the stability of Mo_2C is significantly enhanced.[2] When modified by different coverages of copper, Mo_2C can selectively cleave certain C-O bond, leaving others intact.[3] The concept and feasibility of this work are illustrated using combined theoretical calculations and experimental evaluations over model surfaces and industrial-relevant powder catalysts, which should provide insights into the rational design of efficient carbide-based catalysts for the valorization of biomass-derived oxygenates.

Overall the course of my Ph.D. research, I have developed many experimental and analytical skills, including the operation and tuning of mass spectrometer (MS), high-resolution electron energy loss spectrometer (HREELS), the synthesis and characterization of high surface area catalysts, calculation of reaction energetics using density functional theory (DFT), data analysis with Python, etc. The combination of these techniques allows me to understand the properties of catalysts thoroughly and guide me in catalyst design. I have also spent significant amount of time in maintaining, repairing and upgrading all the equipment in our ultra-high vacuum chambers. These experiences allow me to understand the fundamental principles of the instruments, which help to enhance the accuracy of data measurements and is beneficial to troubleshooting when problems arise. It will also be helpful when I need to establish new labs or install new facilities. In addition to the technical skills, I also developed many interpersonal skills over the years. Within the context of research, I have trained junior students in our lab, and collaborate with research groups in multiple universities and national laboratories to secure funding and publish papers together.[123] Outside of research, I have been the teaching assistant for the Engineering Separation Processes and Chemical Engineering Laboratory courses, which help me learn how to tailor the teaching contents according to students' need and effectively delivery the teaching. My experience as the student president of the Columbia Electrochemical Energy Center and student representative of the Catalysis Club of Metropolitan New York has taught me how to lead and collaborate with our team members to ensure the smooth operation of the organization and to better engage with students and professional members.

Overall, throughout my doctoral study, I have established solid technical and interpersonal skills that can be readily applied to my future position. I would welcome the opportunity to discuss what I can add to and how I can grow with your company.

References

1 Z. Lin, R. Chen, Z. Qu and J. G. Chen, *Green Chem.*, 2018, **20**, 2679–2696.

2 Z. Lin, W. Wan, S. Yao and J. G. Chen, *Appl. Catal. B Environ.*, 2018, **233**, 160–166.

3 W. Wan, S. C. Ammal, Z. Lin, K.-E. You, A. Heyden and J. G. Chen, *Nat. Commun.*, 2018, **9**, 4612.

553340 Investigating Influences of Geometric Factors on a Solar Thermochemical Reactor for Two-Step Carbon Dioxide Splitting Via CFD Models

Solar thermochemical processes utilize inexhaustible solar energy as a thermal driving force to provide heat and motivate reactions, which offer pathways to store solar energy as chemical fuels. Some elevated temperature reactions, like synthesis of syngas and methane reforming, are anticipated in solar reactors which can provide extremely high temperatures. Presently, the dominating considerations in the design of solar reactors are heat and mass transfer with reaction mechanisms. A novel partition cavity-receiver reactor concept is proposed in this paper. In order to provide a longer pathway of interaction between the catalyst and reactants, a partition is introduced in this cavity-receiver reactor. A numerical computational fluid dynamics (CFD) analysis is performed to study the influences of geometric factors (e.g., gap size between partition and bottom, inlets/outlets position, catalyst thickness) under both uniform and concentrated radiant fluxes. A two-step solar thermochemical redox reaction using ceria as a catalyst to split CO_2 is modeled in the partition cavity-receiver reactor to investigate the relationship between geometric factors and reaction rates. Based on the comparisons and analysis of results, optimized geometric factors and corresponding operating conditions are discussed.

553378 Towards Safe and Reliable Eco-Industrial Parks: An Integration and Optimization Approach

The development of Eco-Industrial Parks (EIPs) has drawn much attention due to the growing concern in sustainability. The main idea of an EIP is to form an interdependent production infrastructure that combines chemical plants for integrations of materials, energy, water, utility system, and waste management. However, such highly integrated processes usually result in complex systems with a large number of connections for mass and heat transportation. Thus, a minor incident of one individual plant could lead to a potential decrease in the overall production rate or even complete shutdown of the entire production site. The economic loss due to the event could outweigh the benefits of an integrated system. Safety and reliability considerations should always be included at both design and operation phases. My research focuses on addressing those issues from different aspects within EIPs. The two major areas I have looked into are safety analysis of supply chain network and reliability analysis of water-energy nexus inside industrial parks.

Transportation of chemical plays a central role in supply chain management and multi-plant mass integration through Eco-Industrial Parks. The transported materials often include hazardous materials (HazMat) for which safety is a major concern. My work applies a systematic approach to extract and analyze historical incident data for HazMat to quantify the transportation risk. An optimization framework is then developed to integrate the risk factor into the traditional material allocation problem. The result of my work illustrates a Pareto optimal curve that shows the trade-off between transportation cost and risk, which can guide the decision-making process for optimizing supply chain network within EIPs.

Water-energy nexus analysis is another significant aspect of EIPs. It is one of the promising techniques that can help integrate wastewater into the existing infrastructure and reduce fresh water consumption. I have been working on the development of an optimization framework for the synthesis of reliable direct water recycle network inside industrial parks. My work proposes a statistical model to quantify unit and system reliability, and then integrate it into a mathematical model that represents the direct water recycle network. The model also includes various methods to improve the overall system reliability, such as parallel units and storage tanks. My work results in Pareto optimal solutions that show the trade-off between network cost and reliability. It could help the management to decide the optimal design and operating strategy of the water recycling system to ensure a reliable EIP.

I am a fifth-year PhD candidate from Dr. El-Halwagi's group at the Artie McFerrin Department of Chemical Engineering, Texas A&M University. My research is in the area of process systems engineering, which focuses on process design and integration, mathematical modeling and optimization. I have experience in optimization theory and algorithms, mathematical programming, and supply chain optimization. I have a strong background in process safety and quantitative risk analysis, where I hold the Safety Engineering Certificate from the Mary Kay O'Connor Process Safety Center. I am also interested in data analysis and machine learning. I have performed various analysis techniques and regression studies on extensive data sets in my research. I am proficient in GAMS, R, VBA, Matlab, Aspen Plus, and Python. I received my bachelor's degree from the University of Wisconsin – Madison with double majors in chemical engineering and mathematics.

553465 A Comprehensive Thermodynamic Model for High Salinity Produced Waters - Aqueous Na^{+}-K^{+}-Mg^{2+}-Ca^{2+}-Cl^{-}-so_4^{2-} System

Reuse of high salinity produced waters in oil and gas production is a major concern due to high treatment cost and regulations on disposal in the environment [1]. To support development of novel separation techniques and process innovations, we report development of a comprehensive engineering thermodynamic model for the aqueous hexary system of Na^{+}, K^{+}, Mg^{2+}, Ca^{2+}, Cl^{-}, and SO_4^{2-}, the major ionic species present in high salinity produced waters. Based on the electrolyte NRTL theory [2], the model accurately predicts thermodynamic and phase equilibrium properties with two binary interaction parameters per water-electrolyte pair and electrolyte-electrolyte pair sharing a common ion. The model is validated with the electrolyte concentrations up to salt saturation and temperatures from 273 to 473 K. This paper presents the methodology to identify the binary interaction parameters from the literature data. Furthermore, it presents the model predictions for wide varieties of thermophysical and phase equilibrium properties including salt solubilities for selected binary, ternary, and quaternary subsystems. The model should be a very useful tool in process research, development, and simulation of novel treatment processes for high salinity produced waters.

References

[1] D.L. Shaffer, L.H. Arias Chavez, M. Ben-Sasson, S. R.-V. Castrillón, N.Y. Yip, M. Elimelech, Environ. Sci. Technol. 2013, 47:9569-9583.

[2] Y. Song, C.-C. Chen, "Symmetric Electrolyte Nonrandom Two-Liquid Activity Coefficient Model," Ind. Eng. Chem. Res. 2009, 48:7788-7797.

553493 Nano-Scale Modifications Promote the Success in Industrial Scale: Catalytic Modified Oxygen Carriers for Advanced Fuel Combustion and Gasification System

Abstract

With the existence of severe global warming issue, it is in great need to develop and successfully commercialize clean fuel conversion technologies. Among all the technologies, the chemical looping technologies have multiple applications in fuel combustion as well as methane, coal and biomass reforming. The greatest advantage of chemical looping process is that it can result in a dramatic reduction of capital cost required for the process and improvement of carbon utilization efficiency.

However, the success of the commercialization of the chemical looping system is dependent on oxygen carriers' reactivity and recyclability, whose mechanism requires studies from nano-scale. The core of oxygen carriers' efficiency relies on complex redox reactions involving hydrocarbon molecule absorption and dissociation on metal-oxide based oxygen carrier surfaces, lattice oxygen ion diffusion, oxygen vacancy creation and annihilation at high temperatures. We utilize iron-based materials for oxygen carriers of chemical looping process. Low cost, multiple phases and environmentally-friendly nature of iron oxide leads to the industrial potential of chemical looping utilization. However, the moderate reactivity and selectivity of iron oxide with fuel reaction still impede the success of the commercialization of chemical looping process, where improvement is needed.

Nano-Scale and Lab-Scale Studies:

With a thorough understanding in the mechanism between the reaction of methane and oxygen carriers through density functional theory (DFT) and several years of research with characterization using XRD, XPS, SEM, TEM, and MS. I, with my colleagues:

- Invented catalytic doping method for oxygen carriers to increase the reactivity by 3 times with better recyclability and negligible cost augment
- Created hybrid materials, which increased the selectivity of syngas from 60% to around 95%

I published **7 papers** with my first or leading authorships in this area (3 selected papers are shown below):

- Cobalt Doping Modification for Enhanced Methane Conversion at Low Temperature in Chemical Looping Reforming Systems. *Catalysis Today*, **(in progress). (2019)**
- Enhanced Methane Conversion in Chemical Looping Partial Oxidation Systems Using a Copper Doping Modification. *Applied Catalysis B: Environmental, 235*, **143-149. (2018)**
- Impact of 1% Lanthanum Dopant on Carbonaceous Fuel Redox Reactions with an Iron-Based Oxygen Carrier in Chemical Looping Processes. *ACS Energy Letters, 2*, **70–74. (2016)**

In addition, I took the lead in **2 funded projects** from OH Development Service Agency (ODSA):

- Redox Mechanism Study of Iron-Based Oxygen Carriers in Coal-Direct Chemical Looping Reactions
- Catalytic dopant in modified oxygen carriers for coal-direct chemical looping applications

Bench-Scale and Sub-pilot-Scale Tests:

We invented our new particle based on our nanoscale knowledge and tested their properties in lab-scale. The ultimate goal of oxygen carriers' development is commercialization. By co-operation with the other people in our group, I tested the new particles in bench-scale and sub-pilot scale, where I studied the kinetics of the particle by using machine learning fitting and core-shell model. In order to finish these tasks, I learned Java, Python, C, Matlab and Machine Learning during my spare time. I utilized the knowledge and:

- Tested particles in syngas generation and fuel combustion as the bench unit, gaining a conversion of fuel higher than 98% with 90% syngas selectivity or 99% fuel combustion
- Built an automatic program in the labs to control gas injection without manual handling
- Constructed the program with automatic data analysis and graph depicting, leading to a 10 times faster experimental efficiency

2 papers were published for the bench scale and sub-pilot scale test with my leading authorship (and two papers about kinetic model are in progress):

- High Purity Syngas and Hydrogen Coproduction Using Copper-iron Oxygen Carriers in Chemical Looping Reforming Process. *Applied Energy, 235*, **1415-1426. (2019)**
- A Novel Chemical Looping Partial Oxidation Process for Thermal Oxidation Process for Thermochemical Conversion of Biomass to Syngas. *Applied Energy, 222*, **119-131. (2018)**

Meanwhile, I made great contributions to several **funded projects** from US Department of Energy (DOE) and ODSA:

- Biomass gasification for chemical production using chemical looping techniques
- Chemical looping coal gasification sub-pilot unit demonstration and economics assessment for IGCC application

Pilot-Scale Operation:

Finally, we tested our modified particles in our pilot scale units both in National Carbon Capture Center (NCCC) for Syngas Chemical Looping (SCL) system and Babcock & Wilcox (B&W) for Coal Direct Chemical Looping (CDCL) system. As a member of running the pilot scale unit:

- We achieved **both 1000 hours of steady solid circulation operation** for both unit with **more than 100 hours** steady fuel injection and **95% or higher fuel conversion**
- I led a programming team for lab-scale and pilot-scale data analysis using andvanced MATLAB and Python techniques and skills
- I led the modification and safety inspection (HAZOP, lockout tag-out) process and a professional team for setting up and maintaining gas analyzer for pilot plant operation
- I negotiated and communicated with technical engineers on the operation site, focusing on scheme decision, design modification and schedule planning with my extraordinarily communication skills for industrial cooperation

We published **a paper** for NCCC operation and another paper for B&W unit is undergoing:

- 250 kW_{th} High Pressure Pilot Demonstration of the Syngas Chemical Looping System for High Purity H_2 Production with CO_2 *Applied Energy, 230*, 1660-1672. (2018)

I also made great contributions to these two large **funded projects (each with more than 5 millions dollars)** from National Energy Technology Laboratory (NETL) and DOE:

- Pilot scale operation and testing of syngas chemical looping for hydrogen production

- Commercialization of an atmospheric iron-based coal direct chemical looping process for power production 10 MWE coal direct chemical looping large pilot plant – Pre-front end engineering and design (pre-feed) study

Carbon Materials and Batteries Experiences:

In addition, I'm also experienced in studies of carbon nanomaterials (carbon nanotubes and graphene) and lithium related batteries. I led an undergraduate group and created phosphoric acid modified carbon nanotubes/sulfur composite as cathode material for lithium-sulfur batteries with enhanced lithium storage performance and long-term cycling stability. I published two papers with first and leading authorship in this area:

- Hydrothermal Synthesis of Porous Phosphorus-doped Carbon Nanotubes and Their Use in the Oxygen Reduction Reaction and Lithium-Sulfur Batteries. *New Carbon Materials*, *31*(3), 352–362. (2016) (**27 citations**)
- Strongly Coupled Interfaces between A Heterogeneous Carbon Host and A Sulfur-containing Guest for Highly Stable Lithium-Sulfur Batteries: Mechanistic Insight into Capacity Degradation. *Advanced Materials Interfaces, 1(7).* (2014) (**221 citations**)

Teaching:

Except for research, I am also an excellent teaching assistant in Transport Phenomena and Process Design, Economics, & Strategy class. As a teaching assistant with the responsibilities of helping both the professor and undergraduate students:

- I led a teaching assistant team for homework grading, exam grading, office hour teaching, and feedback collecting from students to professors
- I designed homework, checked the solutions for examination, and made corrections on those errors during the office hours
- I communicated with students about their study plans, research interests, and how to use the course knowledge from their research to their everyday lives, providing them a deeper understanding of the courses

Meanwhile, I am a mentor of two undergraduate students and several Ph. D. students. As a mentor of the undergraduate students:

- I helped them to study experiment skills, such as TGA, sol-gel synthesis, XRD, and simulation skills, such as MATLAB and Aspen Simulation
- I guided them in how to read papers, how to think creatively, how to schedule the experiments, and how to start their own research
- I inspired them to set the goal to finish their undergraduate theses

As a mentor of the graduate students, not only did I help them in areas above, I also:

- assisted them in learning how to run the system during the 1000 hours operation
- instructed them to get familiar with our pilot scale facility and gave them some suggestions and ideas about their research and theses

My studies in nanoscale promote the success of industrial scale, and my industrial scale experiences prepared me a great candidate with the capability of creative thinking from all scales, and thus enabled me to make a big difference for industrial work.

Technical and scientific skills:

- **Experimental skills:** X-ray Diffraction (XRD), Scanning Electron Microscope (SEM) and Focused Ion Beam (FIB), Transmission Electron Microscope (TEM), X-ray Photoelectron

Spectroscopy (XPS), Energy-dispersive X-ray spectroscopy (EDS), Thermogravimetric Analysis (TGA), Mass Spectrometer (MS), Gas Analyzer, FTIR
- **Process design:** Pilot Unit Operation, HAZOP, Sub-pilot Unit Operation, Bench Unit Operation, Fixed Bed Reactor, U-tube Reactor
- **Simulation:** Aspen Plus, Factsage, Density Functional Theory (DFT), Applied Machine Learning, Applied Data Science
- **Programming:** Python, Java, C; Matlab, Pandas, Numpy, Scipy, Sklearn
- **Software:** Microsoft Office, Photoshop, Origin, AutoCAD, Premiere, Dac-Factory

553582 Electrostatically Enhanced Catalytic Phase Transfer Hydrogenation of Acetophenone Under Low External Electric Field

Phase transfer catalysis (PTC) has been extensively studied as a practical method for organic synthesis, as it not only achieves high reactivity and selectivity but also features operational simplicity, and economic and environmentally benign reaction systems. While having various advantages, PTC processes are not as efficient as homogeneous catalysis due to the mass transfer limitation between the two essentially immiscible liquid phases. Additionally, the novel design of phase transfer catalysts is a significant challenge, which limits the potential of PTC for large-scale applications. Here we show that an oriented external electric field (OEEF) can work as phase transfer agent to accelerate the migration of the desired species into a second liquid phase with the absence of phase transfer catalysts. We demonstrated experimentally that an oriented external electric field (OEEF) can act as promoter or inhibitor for the phase transfer hydrogenation of acetophenone depending on its direction. Organic solvents showed differing influences on the reaction attributed to different electrical conductivities. A non-linear dependence of product yield on OEEF potential was also discovered.While further systematic study should be done to obtain insights of the mechanism, this method opens up an opportunity for easy and economic intensification of a wide range of organic reactions with OEEF.

553839 Towards a Sustainable Biorefinery: Biomass for Energy and Biochar for Water Treatment

Towards a Sustainable Biorefinery: Biomass for Energy and Biochar for Water Treatment

Sai Teja Neeli

The Russell School of Chemical Engineering, The University of Tulsa, OK, 74104, United States

Research Interests:

With the increase in energy, there is a growing concern for replacing or at least complementing depleting fossil fuels with alternative sources of energy. Biomass can be regarded as one of the abundant sources of renewable energy for the production of transportation fuels and could be an important part of a more sustainable future energy system. Biomass has a complex structure wherein cellulose, hemicellulose and lignin are intertwined. My research interest lies in how to explore biomass so that even its by-products can be converted to value-added products. In order to be able to transform any biomass derived products, the fundamental science behind the processes have to be unraveled. I am interested to learn the behavior of each of these biomass model compounds to have a better understanding of biomass as a whole. Investigating the model compounds in biomass will reveal which one (s) are responsible for maximum product formation. In my research, I applied hydrothermal water as the medium for the production of biocrude and biochar, a carbon rich by product. Hydrothermal technology has many advantages such as high heat and mass transfer rates, feedstock flexible and no drying step is required. To achieve a sustainable biorefinery, it is necessary to utilize all the products and minimize the waste. Therefore, after liquefying biomass model compounds hydrothermally to obtain biocrude (bio-oil), the biochar left as residue was converted to an adsorbent for the adsorption of heavy metals. Throughout my research, I utilized the fundamentals of a chemical engineering to design and optimize a process while drawing key principles from organic chemistry and materials synthesis to understand the mechanism of product formation.

Research Experience:

During my doctoral studies, I helped my advisor to set up her new research lab. In the process, I ordered and built my experimental set up from scratch and organized the laboratory. I not only used sub- and supercritical water to convert biomass to biocrude and biochar but I also synthesized carbon encapsulated iron nanoparticles (CEINP) derived from biochar. Some of my other duties include maintaining and troubleshooting analytical instruments and writing standard operating procedures for each equipment and experiment. Also, I was selected as a researcher in Oklahoma Catalyst Program which helps graduate students to turn their research into a business. Through this program, I got to learn crucial concepts in entrepreneurship such as intellectual property issues, funding opportunities, identifying customers and understanding a problem.

Below are the details of the research projects I am involved with:

1. *Biofuel production*

During my Ph.D., I studied the effect of a hydrogen donor (calcium formate) on the liquefaction of biomass model compounds. Experiments were carried between 350 and 400 °C in a stainless steel batch reactor. A detailed analysis of the chemical composition of biocrude through GC-MS provided information on the possible reactions in sub- and supercritical water with formate salt. The formate salt improved the biocrude yields and quality (rich in alkylated phenols) appreciably during lignin liquefaction through hydrodeoxygenation reactions. Calcium from the formate salt enhanced the gasification of cellulose in hydrothermal media, and the excess hydrogen improved the quality of biocrude (rich in aromatics) restricting the formation of PAHs. GC-MS results of biocrude from xylan liquefaction with calcium formate showed the presence of phenolic compounds and cyclic ketone. In this study, the presence of calcium formate creates an alkaline environment catalyzing the decomposition of xylan to phenolics and ketones via aldol condensation reactions in sub- and supercritical water. In this work, I proposed reaction mechanisms for the transformation of biomass to biocrude in the presence of hydrogen donor. This reaction not only enabled me to learn analytical techniques like Higher Heating Value (HHV), GC-MS but also made me dig deep into the chemistry of the reactions so that I could postulate new hydrodeoxygenation reactions.

2. *Nanoparticle synthesis*

The second part of research uses biochar, the by product from my first research to produce carbon encapsulated iron nanoparticles (CEINP). I used biochar from all three biomass model compounds for comparison since they are structurally and chemically different. The synthesized catalysts were characterized for physiochemical properties using various analytical techniques like X-ray diffraction (XRD), Fourier Transform Infrared (FT-IR) spectroscopy, Transmission Electron Microscope (TEM), Scanning Electron Microscope (SEM), Brunauer-Emmett-Teller (BET) surface area, Thermogravimetric analysis (TGA). In short, the carbon supported iron nanoparticles preparation includes iron impregnation onto the biochar followed by annealing the mixture at 1000 °C resulting in CEINP. These nanoparticles typically consist of a core-shell structure with iron/iron-carbide core surrounded by graphitic carbon shell. Highly-ordered graphitic layers were observed in the CEINP from cellulose and hemicellulose biochar while disordered graphitic carbon was observed in CEINP from lignin biochar encapsulating the iron nanoparticles. Based on the observed characterization results, I proposed a detailed mechanism explaining the formation of CEINP.

3. *Waste water treatment*

The carbon encapsulated iron nanoparticles were used for the removal of heavy metals from aqueous solutions is tested. Heavy metals are toxic to both human and other living forms posing a serious environmental concern nowadays. United States Environmental Protection Agency (US EPA) has enforced regulations to limit the level of inorganic chemicals in drinking water. Present technologies such as precipitation, membrane filtration and ion-exchange are being employed to remove metal pollutants from water. Unlike these methods, adsorption has proven to be economical and efficient for removing heavy metals. Batch experiments were conducted to test the ability of carbon encapsulated iron nanoparticles for removal of heavy metals (Cr (VI), As (III), Cu (II) and Pb (II)) from aqueous solutions. Higher removal efficiencies (close to 90%) were obtained at pH: 3 for chromium and arsenic while at pH: 7 copper was removed completely. Due to the magnetic property of iron, CEINP can be easily removed from the solutions by applying external magnetic field. Also, these nanoparticles can be potentially reused after heavy metal treatment. Cellulose biochar alone without Fe annealed at 1000 °C was less efficient (<20% efficiency) in removing chromium and arsenic.

Teaching and Mentoring Experience:

Apart from my research career, I have been a Teaching Assistant since the beginning of my graduate program. Occasionally, I was teaching undergraduate level Heat and Mass Transfer courses. Furthermore, I assisted students in performing laboratory experiments related to enzyme kinetics, composition analysis, bomb calorimeter and pump performance. During my graduate study, I had the privilege of mentoring 8 undergraduate students who worked directly with me. I closely mentored the students working on different parts of my research in which they had to apply concepts related to heat and mass transfer, fluid mechanics and reaction engineering to understand the chemical reactions involved. These students went on to present their research at several international conferences and some of them pursued graduate studies. Since I my mentor's first graduate student, I also mentored two graduate students who joined our lab. I trained them on safety, lab practices, equipment handling and conduct reactions.

Service:

As a STEM student, I strongly believe in giving back to the community. I volunteered at the 2015-2018 International Petroleum Environmental Conference (IPEC) organized by The University of Tulsa. At this conference over the years, I moved from being a student volunteer, being responsible for registration to being a session chair. I also volunteered to be a judge for undergraduate student poster competitions at 2016-2017 American Institute of Chemical Engineers (AIChE) conference.

Scholarly Achievements:

- Peer Reviewed Journal articles: **2** published (first authored)
- Conference Presentations: **9**
- Graduate Student Research Grant, The University of Tulsa
- Recipient of Chapman Graduate Scholarship, The University of Tulsa
- Recipient of Phillips 66 R&D Fellowship
- Tube Fitting Installation & Certification, Swagelok, Oklahoma

Selected Peer Reviewed Publications:

1. **Neeli, S.T.**; Shakya, R.; Adhikari, S.; Ramsurn, H., Effect of calcium formate on hydrodeoxygenation of biomass model compounds, ***Energy & Fuels,*** **2019**; DOI: 10.1021/acs.energyfuels.8b04205

2. **Neeli, S.T.**; Ramsurn, H., Synthesis and formation mechanism of iron nanoparticles in graphitized carbon matrix using biochar from biomass model compounds as a support. ***Carbon*** **2018**, *134*, 480-490; DOI: 10.1016/j.carbon.2018.03.079

553697 GCxGC-Qms Analysis of Hexylbenzene Pyrolysis

This work studies the chemistry of hexylbenzene pyrolysis under conditions of 450°C, 75 bar, and < 90 minutes residence time through kinetic modeling using Reaction Mechanism Generator (RMG) and experimentation using high pressure batch reactors analyzed by 2-dimensional gas chromatography with quadrupole mass spectrometry (GCxGC-qMS). The underlying motivation of this work is to understand alkylaromatic reactions in petroleum refining, and characterize the formation of two ring aromatics, likely precursors to coke. This work utilizes GCxGC-qMS to identify pyrolysis products that have two or more aromatic rings; these compounds are categorized into (I) bridged two ring aromatics, (II) fused two ring aromatics, and (III) compounds with more than two rings. Formation mechanisms of each of the categories are discussed, with supporting thermochemistry and kinetics from RMG's database. Pathways discussed in this work can be extrapolated to higher molecular weight aromatics, bridging the understanding between the primary mechanism of alkylaromatic pyrolysis and coke formation.

553776 Development of Cellulosic Aerogel for Oil Spill Treatment

Cellulose is the most abundant organic polymer on earth and is used to manufacture an aerogel for butanol separation from water. It is a polysaccharide consisting of D-glucose and is inherently renewable due to its origin; green plant and bacteria cell walls. Cellulose can be accessed in its pure form from seed hairs of cotton, or in its native composite material with lignin and hemicellulose.

Cellulose aerogels have received academic attention for use as an effective sorbent. Cellulose aerogels have been successfully used for the absorption of oil and dyes which can be owed to its high porosity and very low density.

Cellulose is inherently highly crystalline, and many inter- and intra- molecular hydrogen bonds are present which makes cellulose insoluble in water and most organic solvents.

The synthesis of a hydrogel is initiated when cellulose is dissolved and many solvents have been investigated such as N-methylmorphine-N-oxide (NMMO), lithium chloride/N,N-dimethylacetamide (LiCl/DMAC), NH3/NH4SCN, and NaOH/PEG. NaOH/urea was chosen due to its success as a solvent for cellulose as well as its availability and environmental-friendliness compared to the alternatives. Hydrochloric acid is used for the regeneration of cellulose. The gelation step involves the use of an anti-solvent to extract solidified castings from the previous step and initiate cellulose aggregation. Many different antisolvents can be used and are investigated, namely, acetone, methanol, ethanol, propanol, and sulfuric acid.

The hydrogel can be converted to an aerogel through various techniques with the most common being supercritical CO2 exposure. Freeze drying has also been investigated with successful formation of a cellulose aerogel through lyophilization. Lyophilization is achieved as the vacuum creates an atmosphere that favours sublimation where ice crystals change directly from solid to vapour without passing through a liquid phase. Cellulose is inherently hydrophilic due to the presence of hydroxyl (OH) groups so must be treated to achieve the necessary level of hydrophobicity to be used as an effective sorbent for butanol in water.

553794 Development of Alternative Materials to Replace Precious Metals in Sustainable Catalytic Technologies

Several reactions relevant to the progress of energy, environment and chemical manufacturing sectors involve the use of catalysts containing precious metals. The primary aim of my research is to develop: (a) precious metal-free carbon-based materials for applications in proton exchange membrane (PEM) fuel cells and electrolyzers, electrochemical synthesis of halogens and removal of halide ions from waste water, and (b) mixed metal oxide catalysts for catalytic combustion of low concentration CH_4 emitted from coal mines. My research work, therefore, involves systematic modification of catalyst synthesis procedures along with performing in-situ, ex-situ and operando characterization using a variety of spectroscopic techniques such as X-ray photoelectron spectroscopy, infrared and Raman spectroscopy, X-ray absorption spectroscopy, X-ray diffraction, Mossbauer spectroscopy and elemental analysis in conjunction with catalytic activity measurements to determine structure-property relationships in both electrocatalysis and heterogeneous catalysis.

Catalyst development for PEM fuel cells and electrolyzers:

One of the greatest challenges in electrocatalysis is the commercialization of PEM fuel cell and electrolyzer technology due to high cost and poor bifunctional characteristics of Pt and Ir catalysts required to overcome the slow kinetics of oxygen reduction (ORR) and oxygen evolution (OER) reactions, respectively. My work has established that carbon-based materials are promising electrodes for regenerative PEM fuel cells by demonstrating better bifunctional characteristics as compared to precious metal catalysts [1]. I have developed several approaches to fundamentally investigate the nature of ORR and OER active sites in these materials and used them together with density functional theory calculations to selectively tailor active sites and identified dopants to enhance their catalytic performance by ~22% [2-4].

Carbon-based catalysts for halogen production and water treatment technologies:

The energy requirement in the production of indispensable chemicals such as Cl_2 and Br_2 via electrolysis reduces by ~30% using oxygen depolarized cathode (ODC) technology. I have not only demonstrated the feasibility of using carbon-based catalysts as ODCs in electrocatalytic halogen production processes but have also identified that the versatile nature of active sites in these materials makes them active towards oxidation of halide ions in aqueous solutions [5]. This reaction finds applications ranging from halogen production to water treatment technologies.

Metal oxides for lean methane combustion:

I am also involved in developing mixed metal oxides containing Pd for catalytic combustion of low concentration CH_4 for application in technologies targeted towards reducing green-house gas emissions from coal mines. The aim is to reduce the loading of Pd in the materials while also reducing the temperature of CH_4 combustion, thereby increasing the economic feasibility of this technology and minimizing NO_x emissions. These studies will provide key insights into catalyst chemistry that can be leveraged to increase the efficiency of lean CH_4 combustion technologies.

References

[1] K. Mamtani, D. Jain, D. Dogu, V. Gustin, S. Gunduz, A.C. Co, U.S. Ozkan, Insights into oxygen

reduction reaction (ORR) and oxygen evolution reaction (OER) active sites for nitrogen-doped carbon nanostructures (CNx) in acidic media, Appl. Catal. B-Environ., 220 (2018) 88-97.

[2] K. Mamtani, D. Jain, D. Zemlyanov, G. Celik, J. Luthman, G. Renkes, A.C. Co, U.S. Ozkan, Probing the Oxygen Reduction Reaction Active Sites over Nitrogen-Doped Carbon Nanostructures (CNx) in Acidic Media Using Phosphate Anion, ACS Catalysis, 6 (2016) 7249-7259.

[3] D. Jain, K. Mamtani, V. Gustin, S. Gunduz, G. Celik, I. Waluyo, A. Hunt, A.C. Co, U.S. Ozkan, Enhancement in Oxygen Reduction Reaction Activity of Nitrogen-Doped Carbon Nanostructures in Acidic Media through Chloride-Ion Exposure, ChemElectroChem, 5 (2018) 1966-1975.

[4] Q. Zhang, K. Mamtani, D. Jain, U. Ozkan, A. Asthagiri, CO Poisoning Effects on FeNC and CN x ORR Catalysts: A Combined Experimental–Computational Study, The Journal of Physical Chemistry C, 120 (2016) 15173-15184.

[5] K. Mamtani, D. Jain, A.C. Co, U.S. Ozkan, Investigation of Chloride Poisoning Resistance for Nitrogen-Doped Carbon Nanostructures as Oxygen Depolarized Cathode Catalysts in Acidic Media, Catalysis Letters, 147 (2017) 2903-2909.

553815 Let's Talk: Iiot Enabled Testbeds & Big Data Analytics Applications for Smart-Manufacturing

LET'S TALK: IIoT Enabled Testbeds & Big Data Analytics Applications for Smart-Manufacturing

Abstract:

Candidate summary:

- Creative engineer with 2 years of industrial experience and 4+ years of research experience in **IIoT applications, big data analytics, smart manufacturing, process modelling & optimization** resulting in 4 successful collaborative research projects & 7+ peer-reviewed publications.

- Strong algorithm and programming experience in **MATLAB, python, Linux high performance computing** environments and **parallel computing** evidenced by numerous written public and private codes during execution of research projects. Public codes are available on GitHub (link on poster).

- Recognized leader & Effective communicator demonstrated by 10+ presentations at international / national conferences and through dedicated service in diverse organizations.

Poster summary:

Why IIoT & Big data analytics?

They enables **real-time (remote) monitoring and control, predictive maintenance, etc.** These capabilities help improve plant efficiencies by achieving better control, reducing off spec production, reducing equipment failures and process down time, etc.

What are our contributions?

We've made contributions to all aspects of smart manufacturing. From hardware to software, from data collection, data management, to data analytics. We have demonstrated how to make standard unit operations smart by utilizing noninvasive IIoT sensors, Raspberry Pi microcontrollers, lightweight wireless communication protocols, systems engineering enhanced machine learning algorithms, model based feature selection, data mining etc.

Specific contributions:

1. **IIoT enabled industrial testbeds:**

Every step involved in developing IIoT enabled testbeds will be discussed. We have tested IIoT temperature sensor, vibration/accelerometers sensor (both digital and analog), near IR light sensor, IR camera, video camera etc.

Key discussion points:

- Lab setup, experimental design & IIoT data characteristics & challenges.
- **Predictive modelling framework** built with brief description of **data compilation, data mining, and**

data analysis procedures.

- Novel model based **data filtering** or feature selection approach capable of selecting relevant information for vibration datasets.

- Performance comparison of **hierarchical linear modelling** with **neural networks (ANN & LSTM).**

2. Statistics pattern analysis (SPA) modelling for spectrum data:

We will discuss SPA based modelling approach developed for spectrum data. Developed approach captures more information with less computation ideal for smart-manufacturing applications.

Basic Idea:

- Divide spectrum in different intervals and calculate statistics for each.

- Identify most relevant statistics and use as independent variables.

Scope:

- Tested on 4 public real world NIR, UV/Vis spectrum data from different industries.

- Compared with 4 other linear, non-linear regression approaches

Major Advantages:

- Better prediction accuracy with improved robustness.

- Simpler model with few important variables.

- Incorporates non-linearity.

- Faster training and validation.

3. Statistics pattern analysis (SPA) modelling for batch data:

Next a variant of SPA modelling approach is presented for **semiconductor** batch data. Proposed **variant was proven to have better performance** by comparing its performance with other approaches used in semiconductor industry.

Proficient with the following data analytics and modelling/optimization techniques:

Partial least squares (PLS), Kernel-PLS, Synergy interval partial least squares, Artificial neural networks, Long short term memory (LSTM) neural networks, Binary matrix PLS, Principal component analysis, Time series analysis, Recursive PLS, Kalman filter, LASSO, Bayesian optimization, Gaussian process, Data filtering, Variable selection, Lomb's Algorithm, Fourier analysis, Signal processing.

553835 Carboxylation of Propylene Oxide to Propylene Carbonate: Kinetics and Catalysis

Isaac Asimov said: "Science can amuse and fascinate us all, but it is engineering that changes the world."

Eminent chemical engineers like George Davis, Carl Bosch, and Dermot Manning offer vast proof of Asimov's insight. Their discoveries and methodologies have truly shaped the world we live in, mostly for better but also for worse. For example, growing up in Hyderabad, India, I witnessed first-hand how industrial technologies can harm the environment. A river walk park where I played as a child used to be known for its beauty and boat rides, but now wreaks with a terrible odor from industrial pollution. Such harmful side effects of chemical manufacturing are not intended. Rather, as Paul Anastas of Yale University said, "well-meaning people often do right things but do them wrong."

This is my mission as a chemical engineer: reinvent technologies for a sustainable future. I have always been enthralled by creative innovations and technologies. But what motivates me the most is the opportunity to make a difference in the world.

On my path to become a chemical engineer, I gained experiences as an intern and as a process engineer. My internship at Nuclear fuel complex, India was especially valuable. It gave me the opportunity to interact with plant managers and see how they struggle with the challenges of day-to-day operations. I also interacted with the plant control division, which monitors critical parameters to ensure safety and efficiency, helping to sample and measure the input and output streams from a unit operation as well as the plant. This data was used to increase plant efficiency and also reduce the waste products in the outlet streams. This project boosted my curiosity to explore solutions for the challenges in constructing and automating chemical plants. It also kindled my interest computer-controlled chemical manufacturing.

Working as a process engineer in Larsen & Turbo, India, a multinational E&PC company, exposed me to the global commercialization of engineering technologies and real-time challenges faced in the construction and start-up of a plant. As part of a team of design engineers, I helped develop innovative solutions to tackle various problems and created engineering design packages for several process treatment units and equipment for the oil and gas industry. These tasks enhanced my knowledge of process flow diagrams PFDs and P&IDs, and simulation software packages such as Aspen, Promax, HTRI. One of the projects I worked on was an inter-company collaborative global project where my team was responsible for designing 100 heat exchangers within a short span of a week. This challenging task improved my time management skills, ability to work under pressure and be motivated in adverse situations. I also had the opportunity to interact with safety and financial departments to seek inputs for economical improvements to plant safety via Hazop studies and improved design for plant unit operations. These responsibilities and opportunities not only enhanced my chemical engineering skills but also my communication and team work skills.

During this period, I collaborated with the safety team to develop a report for the air and water emission standards of various countries to prepare our company for potentially stricter emission standards in the future. My role focused on modifying the current process plant simulations for adapting or adopting new cost-effective technologies. This prompted me to sift through the literature to find innovations in catalysis, solvents, and distillation technologies that our company could adopt. I was struck by how catalysis has dramatically influenced many chemical processes, such as hydroformylation, oxidation, hydrogenation, and polymerization, by minimizing or eliminating wasteful byproducts, increasing process efficiency and improving economic viability.

Because of the critical role that catalysis plays in chemical reactions in terms of sustainability, safety and profitability, I became inspired to pursue doctoral research in this field. I chose to work at the University of Kansas (KU), which has renowned experts in the field at KU's Center for Environmentally Beneficial Catalysis (CEBC). This major life decision took me thousands of miles from home to a college town in the heartland of the U.S.A. that could not be more different than my hometown in India.

I am currently in my 4thyear of doctoral study, and my research focuses on developing processes to synthesize industrial chemicals sustainably and safely. My main project addresses the synthesis of cyclic carbonates from epoxides and carbon dioxide in sustainable and economically profitable manner. Since carbon dioxide is a very stable molecule, it is extremely challenging to activate and convert it at ambient to low temperatures and pressures. Therefore, several catalyst systems have been studied to promote activity and selectivity of this reaction. My goal is to develop new catalysts or use solvents for process intensification and to perform kinetics and mechanistic modeling studies to understand the reaction mechanism. To achieve this, I screened several homogeneous and heterogeneous catalysts and polar solvents. For further studies, ammonium-based salts and alcohols as solvents were tested. This led me to explore how carbon dioxide affects these systems as gas-expanded liquids by performing phase equilibrium studies. I also collaborated with computational chemists to better understand the molecular interactions in the gas-expanded liquid phase, and I conducted kinetic studies on these systems for in-depth analysis of the reaction mechanism, which also involved characterization studies with FTIR and UV-VIS.

Working on this project, not only enhanced my technical and teamwork skills; it also helped me to better understand the intricate details and planning required for process development. The University of Kansas recognized my work and awarded me the "Graduate Student Sustainability Leadership Award" in 2018.

Other aspects of my doctoral research have involved synthesizing nanocatalysts with different structures to analyze the effect of structure on catalyst performance. This study expanded my knowledge of several material characterization tools such as SEM, TEM, and XRD. It also gave me the chance to assist with a grant proposal to the National Science Foundation, providing data to support the proposed research in natural gas reforming with mono and bimetallic catalyst systems. I have mentored and trained dozens of undergraduate students. Furthermore, I am also committed to laboratory safety, and have volunteered for two years to serve on the CEBC's Lab Safety Committee where I conducted lab inspections and presented several training sessions on lab safety.

As a graduate student at CEBC, I interact often with industrial professionals affiliated with the center. These experiences have raised my global awareness and peeked my interest in industrial jobs. I also serve as a teaching assistant, assisting with several undergraduate and graduate courses and leading large discussion sessions for 100+ student groups. This broadened my spectrum of interaction with students from multi-disciplinary and multi-cultural backgrounds. These opportunities not only enhanced my problem solving, team player and interpersonal skills but also my mentoring skills.

With my technical background, teaching skills, and interpersonal communication, I am ready to take the next step on my journey to becoming a chemical engineer and making a difference in the world—for the better.

553842 Effects of Anaerobic Environment in Water Quality Downgradient of a Mulch Biowall

The biowall is one of the permeable reactive barrier techniques used in environmental remediation. A biowall is used for the treatment of the chlorinated solvent contaminants in groundwater. It helps to prevent contaminant discharge from a groundwater plume into surface water or other downgradient receptors. Contaminants, such as chlorinated solvents, are treated within the biowall or immediately downgradient of the biowall. A biowall is made up of a carbon source which is organic in nature. This carbon source promotes anaerobic reductive dechlorination of chlorinated contaminants. Organic material that may be used includes wood mulch or waste plant material and liquid organic material such as vegetable oil or other long chain fatty acids. Anaerobic conditions are created when oxygen is consumed. Additionally, if the water contains a moderate to high amount of sulfate, then the anaerobic conditions lead to the formation of sulfides. The objective of this work is to investigate the concentration of total organic carbon (TOC), sulfate and sulfide by the laboratory experiments and modeling. The laboratory experiments consist of a column test conducted to mimic a biowall. This includes sulfide and sulfate measurement using spectrophotometer absorbance after reacting sample with the reagent chemicals. The second part includes modeling to fit the experimental data. The modeling simulation includes solving the set of partial differential equations obtained from mass balance of different chemical species like cellulose, glucose, lactic acid, and changes in the concentration of sulfate and sulfide using MATLAB. The various model constant parameters, such as the initial concentration of glucose, sulfate, reaction rate constant and velocity are assumed during the modeling simulation. Concentration changes for biomaterials and compounds are compared based on time and distance parameters.

553843 Crude Caprolactam Purification Processing

Abstract

Session: Distillation/Separations

Title: Crude Caprolactam Purification Processing

Authors: Fenil Shah, MS Graduate Student (Texas A&M University – Kingsville), fenilshah017@gmail.com

Dr Matthew Alexander, Associate Professor (Texas A&M University – Kingsville), matthew.alexander@tamuk.edu

Topic Description:

Caprolactam [$(CH_2)_5CONH$] is an organic compound which is used to produce tire cords, nylon fibers for textile, carpet and industrial yarn. Every year, approximately 5 million tons of caprolactam is produced by the manufacturers. The main method for manufacturing of caprolactam is a reaction between cyclohexanone (Anone) and hydroxylamine sulfate (HAS), followed by Beckmann rearrangement using oleum. Anone can be produced by oxidation of cyclohexane, and HAS can be produced by oxidation of ammonia to nitrous oxide, followed by hydrogenation in the presence of sulphuric acid. The crude caprolactam which is processed from cyclohexanone and HAS has different types of impurities like water, ammonium sulfate, light impurities (cyclohexanone, aniline and benzene) and heavy impurities (valeramide, cyclohexanone oxime and octahydrophenazine). So, crude caprolactam must be purified before it goes under polymerization to give nylon fibers.

This research work focuses on simulation and optimization of a crude caprolactam purification process using the Aspen Plus simulator. This crude caprolactam purification plant consists of six distillation columns, one extraction tower and one stripping tower. Currently, we have a design specification for a crude caprolactam purification plant of 150 metric tons (MT)/day. When the plant load is increased to 180 MT/day multiple problems are observed. During the higher material throughput, problems observed are extraction tower temperature remains lower than desired, quality of separation in extraction tower decreases, distillation column pressure drop across the packing materials increases, and bottom operating temperatures remain high, which deteriorates the quality of the product. This research includes the identification of problems during higher material throughput, and subsequent design of the columns for higher feed flowrates using Aspen Plus. After optimization of crude caprolactam purification plant, the capital cost evaluation is also performed using CAPCOST software.

Keywords: Crude caprolactam, high-purity caprolactam, distillation of caprolactam

553848 Technology Development for High Surface Area By Cooling Brine.

Technology development for high-surface area salt by cooling brine

Abstract:

Salt (NaCl) is mainly produced by supersaturating salt-water by heating. During heating when salt-concentration overshoots the saturation limit, salt crystals begin to separate out. The process is used in tropical countries to convert sea-water into salt. Salt can also be produced from mined solutions of rock-salt. The salt produced by this method has grains that are fairly cubical and coarse. The method of producing salt by heating salt-water is energy intensive as latent heat of evaporation of water is quite high. Much of the input energy is lost a waste heat to surroundings. An alternate way is by salt crystallizing salt as the dihydrate ($NaCl \cdot 2H2O$) and then warming it to produce salt that has high surface area owing to its much finer grain size. This process requires less energy and is best suited to colder regions. The technology under development uses a stirred vertical crystallizer to precipitate dihydrate crystals which are then rewarmed to yield salt. This salt has potential in wellness and food industry owing to its adsorption properties and high dissolution rate compared to normal salt. An alternate crystallizer design under development increases compactness and energy efficiency even further by incorporating features of a heat exchanger as an essential part of its design.

553857 Controlling and Optimizing Activity/Stability Balance in Electrocatalytic Materials

Electrocatalysis plays a key role in the energy conversion processes central to several renewable energy technologies that have been developed to reduce our reliance on fossil fuels[1]. The biggest challenge in developing electrochemical energy conversion technologies, including fuel cells and water electrolyzers, is the development of catalysts that are active, selective, and durable. In the design of efficient electrocatalysts, however, catalyst stability is too often overshadowed by the pursuit of ever increasing activities. By studying the relationships between electrocatalytic activity and stability, we aim to identify the relevant mechanisms of these competing performance metrics and develop insight to aid the optimization of a sufficient balance for practical application and device integration.

The development of efficient and stable catalysts for the cathodic oxygen reduction reaction (ORR) is one of the greatest challenges for the exploitation of polymer electrolyte membrane (PEM) fuel cells[2]. After electrochemical dealloying, the open, bicontinuous, three-dimensional nanoporous nanoparticle electrocatalysts exhibit dramatically enhanced electrocatalytic properties[3]. Here, using a combination of in-situ and ex-situ experimental techniques, we studied the instability of this nanoporous materials which can lose electrochemically active surface area (ECSA) through surface smoothening driven coarsening[4]. With a better understanding of the interplay between nanoporous structure coarsening and transition metal loss, we have developed strategies to mitigate coarsening and improve catalyst stability through slowing electrochemically enhanced surface diffusion and limiting step edge movement. The first approach uses ionic liquids (IL) strategically placed at the metal/electrolyte interface to limit electrochemically enhanced surface diffusion. Taking advantage of the free volume within the nanoporous nanoparticles and creating a composite catalyst architecture through the incorporation of [MTBD][beti] IL, in addition to the improved oxygen reduction reaction kinetics by chemically biasing reactants to the surface and changing the degree of solvation/stabilization of adsorbed intermediates[5], significant improvements in retention of ECSA during accelerated durability testing is observed. The interfacial IL acts to limit the charge dependent formation of a surface metal/electrolyte anion complex which is responsible for the potential dependent enhancement in surface diffusion. A second approach involves the minimization of morphology evolution by impeding step edge movement through the use of foreign adsorbates on the surface. We show that partial monolayer decoration of np-NiPt with Ir, possessing a significantly lower rate of surface diffusion than Pt, acts to pin step edges and results in significant enhancement in catalyst durability as measured by ECSA and ORR activity retention[4]. With these strategies we will show how more detailed insight into the atomic processes that govern electrocatalytic material instability can begin to break the inverse correlation between activity and durability.

As an ideal clean energy carrier, hydrogen can be produced through PEM electrolysis by generating H_2 from H_2O. However, noble platinum group metals (PGM) remain the most common electrocatalysts fnor hydrogen evolution reaction (HER). The expense and scarcity of these catalytic materials obstruct the wide-spread adoption of electrochemical fuel generation technologies[6]. Sulfur based transition metal dichalcogenides (TMDs) have emerged as a promising alternative to PGM HER electrocatalysts as they are abundant, inexpensive, and exhibit a low HER overpotential in acidic environment[7]. Here we present a systemic assessment of the compositional dependent HER activity and stability for Co-based mixed chalcogen, CoS_xSe_{2-x}, pyrite TMDs[8]. We observe a decrease in HER activity from the single chalcogen TMDs, CoS_2 and $CoSe_2$, to mixed chalcogen TMDs. This observed compositional trend in HER activity can be explained by the unique combination of compositional dependent hydrogen adsorption free energy (ΔG_{Had}) and bulk resistivity/conductivity of the pyrite TMD. With near thermoneutral ΔG_{Had}, the highest resistivity among the compositions was observed. The following increase in Se content leads to the decrease in HER activity due to a steady movement away from optimal ΔG_{Had}. As the composition approaches $CoSe_2$, however, it is observed that HER activity again increases at higher Se contents, with $CoSe_2$ exhibiting similar activity to CoS_2. This highlights the convolution of ΔG_{Had} and material conductivity in determining the HER activity. Furthermore, through stability tests under constant potential HER electrolysis, Se-rich Co-based pyrite TMDs are found to be more durable than S-rich samples. Therefore, with an HER activity matching that of CoS_2, but with a dramatic improvement in stability, $CoSe_2$ breaks away from the traditional inverse activity/stability relationship and represents a promising material for non-PGM HER electrocatalysis in acidic based PEM electrolyzers.

1. Seh, Z. W. et al., *Science* **355** (2017) eaad4998.
2. Stamenkovic, V. R. et al., *Nature Material* **6** (2007) 241–247.
3. Snyder, J. et al. *Am. Chem. Soc.* **134** (2012) 8633-8645.
4. Li, Y. et al., *ACS Catal.,* **7** (2017) 7995-8005.
5. Snyder, J. et al., *Nature Mater.* **9** (2010) 904-907.
6. Wang, J. et al. *Mater.* **28,** (2016) 215-230.
7. Chhowalla, H. et al. *Nature Chemistry* **5,** (2013) 263-275.
8. Li, Y. et al., *Catal.,* **366** (2018) 50-60.

554583 Production Challenges of Gas Hydrates in Flowlines

Hydrates are a common challenge in the oil and gas production. They are solids forming from small gas and water molecules. Hydrate formation usually requires high pressure and low temperature. The hydrate formation is undirable as it can lead to production interuption and financial loss. The production systems, on the other hand, are always complex with different sections and pipe networks. Different parts of the hydracarbon flowlines are found to face different hydrate challenges.

One common component of the production systems is the gas-filled sections with no through-flow, so called deadlegs. The fluids within the deadlegs are relatively stagnant comparing to the main flowline. If in a cold environment and not properly insulated, the fluids within a deadleg can become so cold that hydrates can form. The field operation showed that deadlegs often encounter hydrate deposition but the understanding was very limited. The risks and the mechanism had not been well studied or publicly available. Therefore, we collaborated with Equior to conduct a systematic study to understand the hydrate deposition in deadlegs. We characterized the hydrate deposition process, quantified the effect of different variables, and modeled the process using transport equations.

The main flowline with flowing fluids can also encounter hydrate problems. However, current experimental setups for hydrates are either very costly to generate flow (e.g. flowloops) or cannot generate flow (e.g. autoclaves). Therefore, we collaborated with Total and DeepStar and developed a new bench-scale setup, named rock-flow cell, to generate flow and study hydrate phenomena under flowing conditions. We characterized the capability of the system by testing different water cuts and liquid loadings. The visual observation from the radial and the axial directions provided rich information to improve the understanding in the hydrate formation process. We also utilized the system to evaluate the performances of various anti-agglomerants. The setup provided satisfactory results to rank the chemicals with the direction observation.

In summary, we have collaborated the industry partners to spend systematic efforts to understand the hydrate problems in flowlines. Our achievement has improved the understanding in the hydrate flow assurance challenges and helped determine the hydrate management strategies.

553804 Pi Electron Cloud Mediated Separation of Aromatics Using Supported Ionic Liquid Membrane (SILM) Having Antibacterial Activity

Pi Electron Cloud Mediated Separation of Aromatics Using Supported Ionic Liquid Membrane (SILM) Having Antibacterial Activity

Mohanad Kamaz[a], Mahmood Jebur[a], Arijit Sengupta[a], Xianhong Qian[b], Ranil Wickramasinghe[a]

[a]Ralph E Martin Department of Chemical Engineering, University of Arkansas, Fayetteville, AR 72701, United States

[c]Department of Biomedical Engineering, University of Arkansas, Fayetteville, AR 72701, USA

Supported ionic liquid membranes (SILM) using imidazolium based ionic liquids and hydrophobic PTFE base membrane were used for the π electron mediated separation of aromatic compounds including dyes in aqueous and non-aqueous medium. The FTIR of SILMs showed the signature of the functionalities for both base membrane and the imidazolium moiety. A significant reduction in contact angle was observed for SILMs compared to the base membrane indicating the enhancement in surface hydrophilicity of the membrane, while considerable changes zeta potential was measured at different pH indicating the modification in surface charge. The SEM image confirmed the pore filling of base membrane with ionic liquids. The stability of the SILMs was studied using hexane, water and DMF solvents. Tuning the anionic moieties, ionic liquids were utilized for aqueous and non-aqueous application through SILMs. For non-aqueous applications, the trend in separation efficiency was found to be divinylbenzene > styrene > toluene, which was attributed to the more π electron cloud density on extended conjugation leading to favorable interaction with ionic liquid influencing its transfer through SILMs. Preferential separation of Congo red (CR) and Ramazol Brilliant Blue R (RBBR) was achieved in water based application of the SILMs.

The ionic liquids were having affinity for the π electron cloud. This π electron cloud density for toluene is lower than styrene followed by divinyl benzene, due to the presence of additional double bonds in conjugations with the aromatic π electron cloud. Due to this extended conjugation, divinyl benzene interacted strongly with the ionic liquid sitting inside pores. This might be responsible for bringing the selectivity in the separation for the SILMs. The lower transport in case of allyl ionic liquid might be attributed to the presence of electron withdrawing sp^2 hybridized carbon of the ally group resulting reduction in the electron density over imidazolium ring. On the other hand, for choloro and bromo ionic liquids, butyl and hexyl groups were present in the side chain of the imidazolium ring. The (+) inductive effect of the alkyl group containing sp^3 hybridized carbon atom pushed the electron density towards imidazolium ring. More electron cloud on the imidazolium ring, it will be more favorable for the π electron interaction resulting enhancement in the separation efficiency.

Similar investigation was also carried out to understand the selectivity in separation of different dyes using SILMs with water insoluble ionic liquids. In this case three different dyes CR, EBT and RBBR were used. The SILMs were found to be more selective for CR. In case of hexyl imidazolium ionic liquid, almost 60 % CR was transferred to the receiver side, while for allyl imidazolium ionic liquid % transport of CR reached around 50 % after two hours. For the other dyes RBBR and EBT, the % transport through both the SILMs were found to be less than 5 %. Though all these dyes have aromatic moieties, their complicated structures with different functionalities and stereo-chemical arrangement might lead to different overall interaction beside electron cloud resulting in different selectivity through SILMs.

The stability of the supported liquid membranes is of major concerns. Therefore, it is interesting to investigate the stability of the SILMs. In view of this SILMs were allowed to be equilibrated in presence of water, DMF and hexane and at different time interval the conductivity of the medium were measured. For SILMs with water soluble ionic liquid, the conductivity of the medium was found to be modified in presence of water and DMF. In presence of water, the water soluble ionic liquids came out of the membrane pores resulting in modifying the supernatant conductivity. DMF being polar in nature, certainly interacts with the ionic liquid moieties in the membrane due to electrostatic interaction. As result, ionic liquid might get dissolved in DMF. This also led to changes in conductivity in DMF, while in case of hexane, up to 60 days, there is no change in the conductivity of the medium indicating the stability of the SILMs. Using these equilibrated membranes (30 days' equilibration), the separation of toluene, styrene and divinyl benzenes were carried out. There modification in separation efficiency were found to be within 10–15 %. These investigations revealed the high stability of SILMs with water soluble ionic liquid in hexane medium.

The antimicrobial activity of these SILMs were investigated using gram-positive bacteria (**staphylococcus aureus** and **pseudomonas aeruginosa)**. More the number of carbon atom on the side chain of the imidazolium ring, more antimicrobial activity was observed. For *staphylococcus aureus*, the L/D ratio for the virgin membrane was kept 52. Over the time frame of 24 hours, the virgin membrane did not show any appreciable modification in the ratio, though some fluctuation is noticed. In case of SILM, a drastic decrease in L/D values were observed for all the ionic liquids upto a definite time (3 hours for bromohexane and chlorobutane ionic liquids; 5 hours for allylbromide ionic liquid), followed by almost constant values for L/D upto 24 hours. This investigating clearly demonstrates the antibacterial activity of SILMs compared to the virgin membrane. The antibacterial activity of allylbromide SILM was found to be slightly lower than that of chlorobutane or bromohexane SILMs. The end point was taken as 24 hours, because beyond that, there might be some appreciable modification in the L/D ratio for the original bacteria due to their life cycle.

Similar study was also carried out with another gram positive bacteria, *pseudomonas aeruginosa* starting with L/D ratio ~ 56 by monitoring the L/D ratio using luminescent spectra for 24 hours with a regular time interval. The trends were found to be similar to *staphylococcus aureus*. Though, this method is very convincing for evaluating the antibacterial activity, still it has some limitations. The bacteria having compromised membranes may be able to recover and reproduce but this method identify them as 'dead' bacteria. Similarly, some bacteria may have intact cell membrane but unable to reproduce in nutrient medium, are assigned as 'live' bacteria. Moreover, since imidazolium moiety is also able to give fluorescence, it may interfere in the determination of fluorescent intensity of the dyes indicating live and dead bacteria. Therefore, additional non fluorescent based technique should also be used to confirm the antimicrobial activity of SILMs.

The quantitative antibacterial analysis was carried out in terms of colony forming units per mL for bacteria. Large CFU/mL values, 2.82 E+09 and 3.5 E+09 were taken as the starting point for *staphylococcus aureus* and*pseudomonas aeruginosa*, respectively. Within 24 hours, there was almost no change in the CFU/mL values for both the bacteria for virgin membrane, while significant reduction in CFU values were observed for SILMs. After 5 hours of contact with the membranes the CFU/mL values become 2.82E+09, 4.4E+06, 1.10E+03 and 1.2E+04 for virgin membrane, SILMs with allylbromide, bromohexane and chlorobutane ionic liquids, respectively; whereas after 24 hours CFU/mL became 2.80 E+09, 9.01E+04, 2.10E+0 and 2.04E+01. The investigation clearly demonstrates, not only the antibacterial activity of SILMs, but also the trend of allylbromide < cholorobutane < bromohexane. This trend was found to be similar than that of earlier investigation. More the number of the carbon atom on the side chain of the imidazolium ring more pronounced antibacterial activity was observed. The quantitative antibacterial analysis was carried out in terms of colony forming units per mL for bacteria. Large CFU/mL values, 2.82

E+09 and 3.5 E+09 were taken as the starting point for *staphylococcus aureus* and *pseudomonas aeruginosa*, respectively. Within 24 hours, there was almost no change in the CFU/mL values for both the bacteria for virgin membrane, while significant reduction in CFU values were observed for SIL membranes. After 5 hours of contact with the membranes the CFU/mL values become 2.82E+09, 4.4E+06, 1.10E+03 and 1.2E+04 for virgin membrane, SILMs with allylbromide, bromohexane and chlorobutane ionic liquids, respectively; whereas after 24 hours CFU/mL became 2.80 E+09, 9.01E+04, 2.10E+0 and 2.04E+01. The investigation clearly demonstrates, not only the antibacterial activity of SILMs, but also the favorable transport trend of allylbromide < cholorobutane < bromohexane.

557297 Inferential Control of Pilot-Plant Distillation Column Using Neural Network Based Soft Sensor

The development of an inferential soft sensor for pilot-plant distillation column of ethanol-water mixture using neural network (NN) method has been investigated in this work. The lags between the input variables and the output variable vary due to changes in operating conditions. An inferential sensor that can infer the composition of ethanol at the top product using time lags for the input variables and varied first-order time constant lags with the output variable is developed. Based on this soft sensor, an inferential PI controller has been developed. The gain scheduling for this controller has been found necessary to get a good performance.

In summary, a high accuracy soft sensor for the ethanol composition of the top distillation product has been developed and validated. Based on this soft sensor and inferential PI controller, Model Predictive Controller (MPC) for this pilot-plant column will be developed in future work.

545276 Designing Industry 4.0 Projects for Process Safety

Designing Industry 4.0 projects for process safety requires understanding the wide spectrum of cyber/physical technologies. This includes Advanced Process Control, Artificial Intelligence, Neural Networks, Digital Twins, Machine Learning, Robots /Cobots, and Operational Guidance Systems. Analytics is a major part of this, but which techniques are most appropriate and how do you best ensure success? Developing a strategy to integrate these technologies involves a process which incorporates goals and a methodology for implementation.

Industry 4.0 is well suited for process safety, and there are key benefits for determining the order of deployment. This may vary from company to company depending on goals and based on your company's maturity/capabilities. Finally, how do you best scale analytic deployments across the enterprise? This talk will focus on these issues and the factors to consider when developing an overall strategy.

Jim Petrusich is the co-author of The Industry 4.0 for Process Safety Handbook *(October 2017).* Northwest Analytics is the winner of the 2018 *Frost and Sullivan* Industry 4.0 Award.

550630 Using Industry 4.0 to Create a Pathway to Operational Excellence

I think we all recognize there is journey to get to Industry 4.0 or Refining 4.0. How do we get to the smart refinery? The smart refinery includes the Internet of Things, Cloud computing, cyber-physical process, and Analytics. Teradata is working with companies on their journey to smart refineries by helping companies with their analytics through IoT and cloud computing at scale of machine learning algorithms. Teradata is providing companies with access to the power of analytics by combining their sensor data and ERP data at granular detail level to gain insights on how to improve costs and increase gross margin in their reliability and optimization processes. We will discuss the methods we apply to to bridge the OT (Operational Technology) and IT(Information Technology) worlds together to have impacts on maintenance programs, energy optimization, and operations. We will also discuss how data integration around applications, data silos , ERP systems have to helping companies not only gain insights but also have financial impact to the bottom line at scale for multiple refineries.

554383 *Accelerating Product Innovation at Dow through Multivariate Modeling* (Poster)

In the era of Industry 4.0, manufacturers such as Dow are focused on gaining more value from their historical data than ever before. ProSensus has distinguished itself as a trusted, global leader in Big Data analytics by helping many Fortune 500 companies across numerous industries use their data to innovate for the future.

ProSensus has recently worked with Dow to accelerate innovation on one of their key product lines – silicone antifoams. For product development applications such as this, ProSensus combines powerful multi-block latent variable modeling with constrained optimization. This approach allows clients to simultaneously optimize the selection of raw materials, recipe formulations, manufacturing conditions and costs to reach targeted product performance properties while adhering to custom constraints. A model-based approach allows Dow to rapidly develop and scale up custom silicone antifoam formulations for the dynamic technical and regulatory needs of customers in the pulp and paper, food and beverage, wastewater treatment, metal working, and other industries.

This presentation will examine the approach taken (data assembly, multivariate modeling, model validation, and optimization), challenges encountered, and results obtained. Perspectives from both ProSensus and Dow will be included.

554363 Reinforcement Learning (RL) Based Data-Driven Decision Support System (Poster)

Big data has transformed from a buzzword to a real value creator. In recent years there has been a rising interest in data analytics in the energy industry. The industry and academia have made significant progress in big data analytics. However, there is still a need to use the plant data to gain insights and address the issues related to abnormal situation management. An abnormal situation is defined as "a disturbance or series of disturbances in a process that causes plant operations to deviate from their normal operating state." The disturbance can be a minor near-miss to a catastrophic incident which may cause production losses, environmental damage, injuries or sometimes fatalities if not managed properly. During an abnormal situation the operator is expected to take appropriate action to bring back the process to a normal state. Diagnose and mitigation actions depend on the operator's experience, process complexity and stress levels of the operator.

This paper outlines a systematic framework for application of Reinforcement Learning (RL) during process operations and abnormal situation management. A novel method using Reinforcement Learning is established to address decision making challenges and advise operators during abnormal situations. The system is designed on the expert knowledge from operators and engineers, the knowledge available from previous operation experiences and process plant information. The application of the proposed method can be expanded by connecting the system to Operator Training Simulators (OTS) and train operators for abnormal situation management.

A sequence to sequence time series forecasting model for batch processes with seasonality

Shu Xu, Mark Nixon

Emerson Automation Solutions

1. Introduction

Modern chemical manufacturing processes generally fall into three categories: continuous, batch and semi-batch. Unlike its continuous counterpart where the raw materials are constantly fed into the system and processed into an intermediate or final product continuously, a batch process receives and processes the feedstock by a series of discrete steps. Such a manufacturing process has been widely used in semiconductor, pharmaceutical, specialty chemical and other industries. One example is the fermentation process used in the brewing industry. Because the whole process, consisting of many operating steps, can hardly be operated in a steady state, it poses tough challenges for both equipment and operators to assure product quality over many batch runs. One specific challenge is related to seasonality which stems from equipment aging[1], periodic maintenance[2], catalyst decay [3] and so on. Typical examples of such a behavior involve activity coefficients of reactor catalysts and fouling factors in heat exchangers[4]. Process engineers are challenged to forecast such a seasonality lurking in the system so that they can become proactive in identifying and preventing potential problems.

In this paper, we built a sequence-to-sequence model to forecast the seasonality simulated in the benchmark Penicillin bio-reactor process. The results obtained by such a sequence-to-sequence model are compared with those generated from a linear partial least square (PLS) method and a nonlinear support vector machine (SVM) regression method.

The remainder of this paper is organized as follows. Section 2 briefly introduces methodology information regarding time series forecasting, long short-memory (LSTMs) networks, and an associated sequence-to-sequence model. Section 3 provides case study details including the benchmark Penicillin bio-reactor simulator and simulation conditions. Section 4 summarizes results obtained in the case study. We conclude the paper and discuss future works in Section 5.

2. Methods

2.1. *Time series forecasting*

During batch operations, abundant informative data can be collected in the form of time series samples with regular sampling periods [5]. As a result, forecasting methods in time series analysis, which take advantage of dependence between observations in the time dimension to predict the future, can be applied in process seasonality forecasting [6]. For example, classical methods include autoregression, autoregressive integrated moving average (ARIMA), simple exponential smoothing (SES), etc. In addition, time series forecasting can also be reframed as a supervised learning problem. By reframing the time series data via a lag method, various linear and nonlinear supervised learning algorithms can be used, including linear regression, support vector machine, neural networks, or a sequence-to-sequence model with long short-term memory (LSTM) networks as discussed in this paper.

2.2. LSTM network

Long short-term memory (LSTM) networks belong to the category of recurrent neural networks (RNNs, see Figure 1) where loops and chains exist to allow information to persist. In Figure 1, an input x_t is fed into a regular neural network block A which produces an output h_t. An RNN distinguishes itself from its regular counterpart in that each neural network unit in the RNN can pass information either to itself in a loop or to a successor to form a chain. Although a normal RNN can track the short-term dependence between observations in the time series, it struggles to learn the "long-term" connections [7]. To overcome such a challenge, the LSTM network is proposed[8] (see Figure 2).

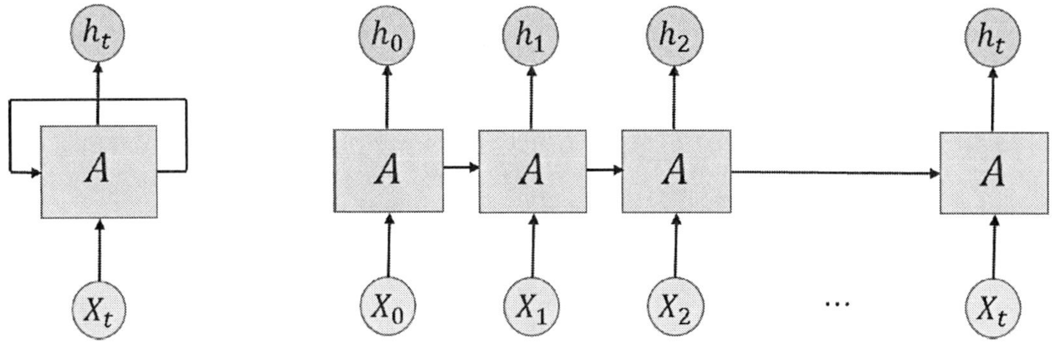

(a) An RNN with loop (b) An unrolled RNN

Figure 1 Recurrent neural network demonstration

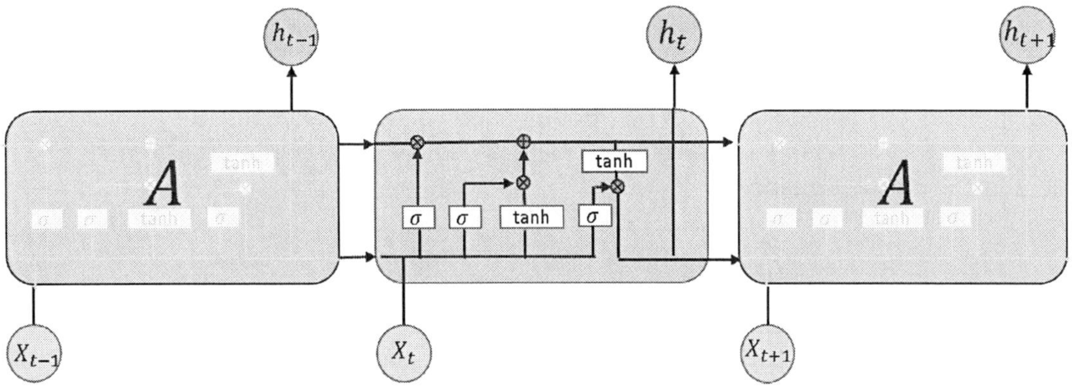

Figure 2 LSTM network demonstration

Although the general form of the LSTM network is akin to that of a regular RNN, each cell contains four neural network layers that interact in a way that enables them to remember information for long periods of time. A detailed description of the structure of each cell follow.

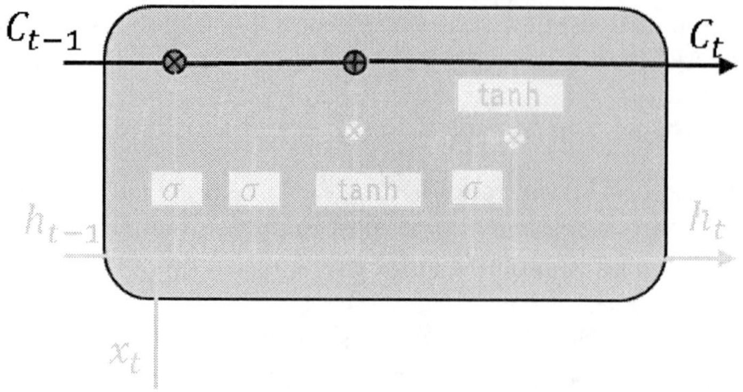

Figure 3 Cell state in an LSTM cell

As illustrated in Figure 3, a horizontal line running across the top of the cell with only linear pointwise operation indicates a potentially free flow of information without any cell status changes. In addition, the amount of cell status information gain or loss is controlled by the following gates in Figure 4, which are comprised of a sigmoid neural net layer and a pointwise multiplication operation. Because the sigmoid layer outputs a number between 0 and 1, it regulates how much information can come through. While a value of 0 indicates closing the gate, a value of 1 means letting the information flow through freely.

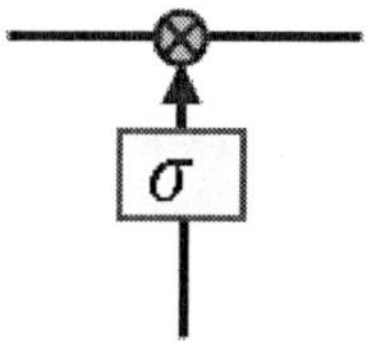

Figure 4 Gate in an LSTM cell

Figures 5 through 8 provide a step-by-step walk-through of the four-layer structure inside each LSTM cell. As shown in Figure 5, the first layer, named the "forget gate layer", decides whether to keep the previous cell state C_{t-1} or not. For example, if the forgetting factor f_t calculated by gate Equation (1) equals 0, the previous cell state is completed discarded (or "forgotten") after an element-wise multiplication.

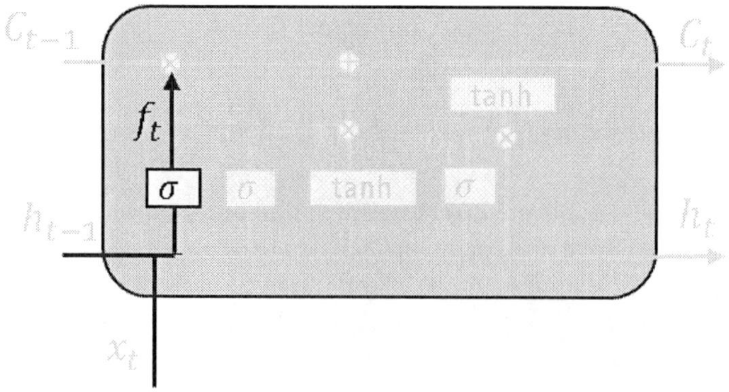

Figure 5 Forget gate layer in an LSTM cell

$$f_t = \sigma\left(W_f \cdot [h_{t-1}, x_t] + b_f\right) \quad (1)$$

The next state update layer consists of two sub-layers. Out of these layers, the "input gate layer" decides whether we need to update the cell state with the new value produced by the "new value layer", as shown in Figure 6. The associated gate and value equations are listed in Equations (2) and (3) respectively.

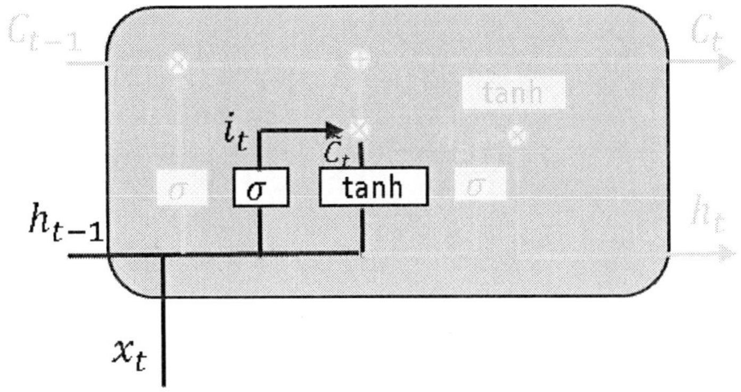

Figure 6 State update layer in an LSTM cell

$$i_t = \sigma(W_i \cdot [h_{t-1}, x_t] + b_i) \quad (2)$$

$$\tilde{C}_t = tanh(W_C \cdot [h_{t-1}, x_t] + b_C) \quad (3)$$

Now since we have both the previous cell state C_{t-1} and the new state update value $\tilde{C}_t$, the new cell state C_t can be calculated by Equation (4). Figure 7 shows the state update.

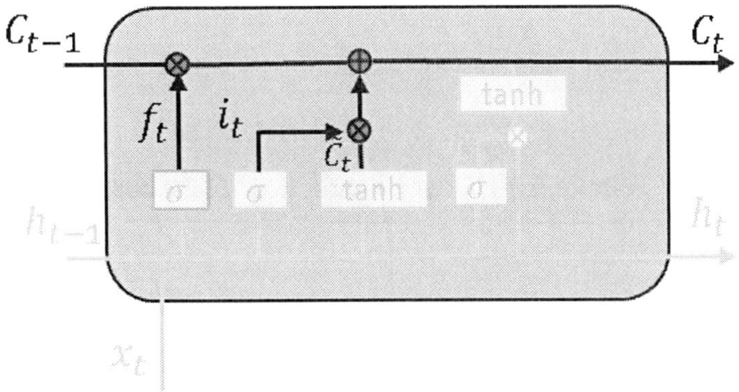

Figure 7 State update operation in an LSTM cell

As shown in Equation (4), the new state C_t equals the sum of an old state C_{t-1} multiplied by the forgetting factor f_t, and a new state update $\tilde{C}_t$ scaled by a factor i_t.

$$C_t = f_t * C_{t-1} + i_t * \tilde{C}_t \quad (4)$$

Finally, the output of the cell is decided by the output layer shown in Figure 8, including a sigmoid gate which decides whether we need to deliver a scaled version of the cell state C_t by "tanh" function, i.e., the h_t. Associated equations are shown in Equations (5) and (6).

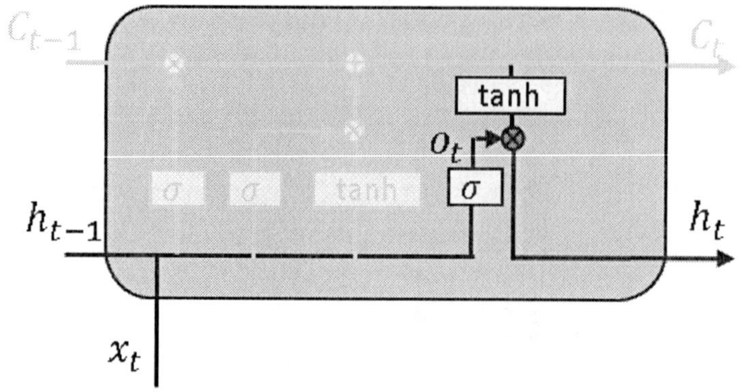

Figure 8 Output layer in an LSTM cell

$$o_t = \sigma(W_o \cdot [h_{t-1}, x_t] + b_o) \quad (5)$$

$$h_t = o_t * tanh(C_t) \quad (6)$$

In our study, a sequence to sequence (seq2seq) model with LSTM cells is used for time series forecasting. The model includes an encoder converting an input sequence X_t to corresponding state vectors which contain transformed context for each input point x_t, and a decoder which takes context vectors and internal hidden states as inputs to generate an output sequence Y_t.

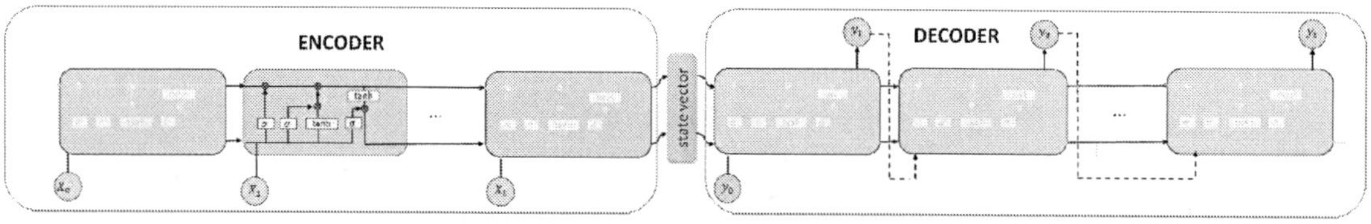

Figure 9 A seq2seq model with LSTM cells

3. Case study

3.1. *The Benchmark Penicillin Bio-reactor simulator*

A lumped-parameter dynamic batch-fed Penicillin model developed by Birol et al. [9] is used as the basis for the current study. The original MATLAB code for the simulation program can be downloaded from Dr. Cinar's website (www.chee.iit.edu/~cinar/software.htm), and modifications have been made on the key kinetic parameter to simulate the batch seasonality. The main state variables included in this simulation are: reactor temperature, biomass volume, pH, biomass concentration, substrate concentration, penicillin concentration, dissolved oxygen concentration, and carbon dioxide concentration. The full model is provided in the Appendix. The ODE23t stiff solver provided in MATLAB is used to simulate the process.

The general process flow diagram of the penicillin reactor is shown in Figure 10. In the initial growth phase, the fermenter is initiated with a small volume of bacteria culture and substrate (glucose and oxygen). The reactor temperature is kept at 298 K (room temperature) to initiate the growth of the Penicillin-producing bacteria. After approximately 50 hours, the initial substrate is depleted; the reactor then enters the penicillin production phase. In this phase, the fermenter is operated in semi-batch mode. Continuous flows of substrate (through the substrate tank) and oxygen (through the aerator in the bottom of the fermenter) promote the production of penicillin, carbon dioxide and other by-products. Furthermore, two PID controllers regulate the acidity level (pH ~5) and temperature (298 K) of the fermenter to promote penicillin production.

Figure 11 plots the resulting trajectory of a single batch. The sampling frequency is set to once every 30 minutes, mimicking that of real plant data. Output variables of the simulation include agitator power, aeration rate, substrate flow rate, cooling water flow rate, heat generation. These variables are either manipulated inputs, or measured variables of the PID control loop.

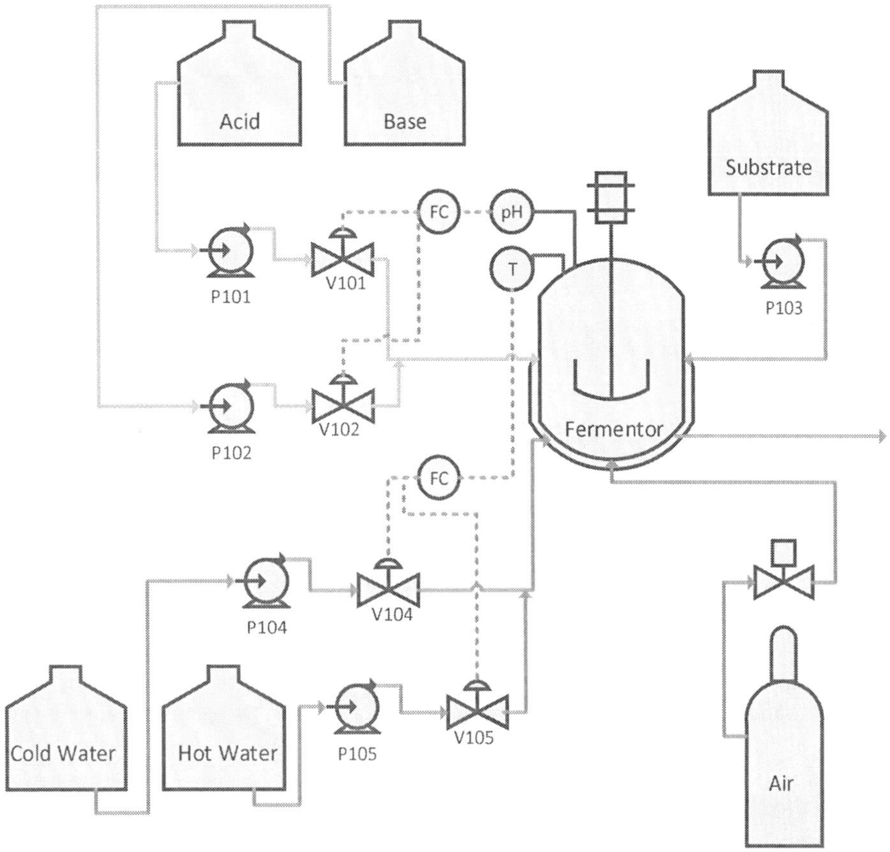

Figure 10 Process flow diagram of the batch-fed penicillin process

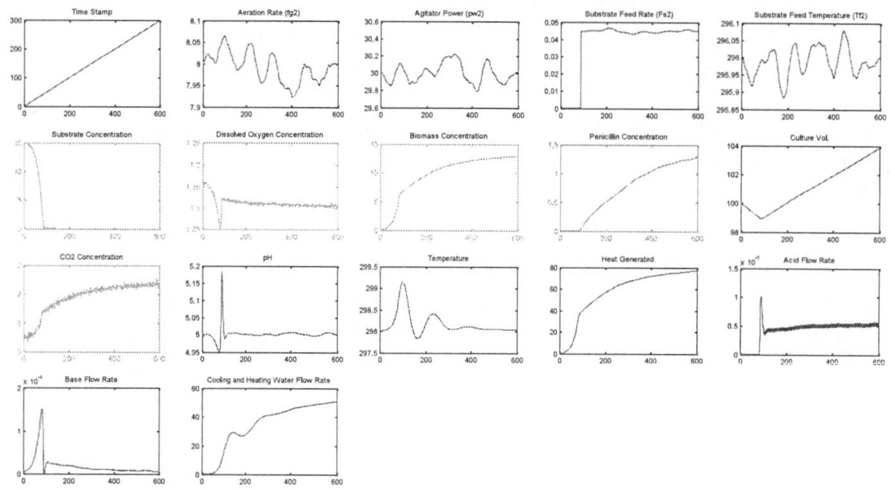

Figure 11 Example batch trajectories from a simulation run of the penicillin reactor

3.2. Simulation Conditions

To introduce process seasonality, the specific growth rate $\boldsymbol{\mu}$ in the biomass growth equation (see Equation (7)) is configured to follow a first-order exponential decay with a time constant set to 150 batches. The decay of the specific growth rate $\boldsymbol{\mu}$ starts from the nominal value of 0.092 and ends with the terminal value of 0.048. As shown in Figure 12, scheduled maintenance (which occurs once every 100 batch cycles) of the reactor resets the specific growth rate back to its nominal value.

$$\frac{dX}{dt} = \mu X - \frac{X}{V}\frac{dV}{dt} \tag{7}$$

Table 1 lists the simulation conditions for our current study. A total of 600 batches with seasonality corresponding to six cycles in Figure 12 are simulated, out of which 400 batches are used for training, and the remaining 200 are used for testing. All the data goes through batch-wise unfolding as shown in Figure 13, where 3D batch data are flattened batch-wise into a 2D matrix. After unfolding, data from the ith batch are squeezed into the ith row of the matrix, and such a row is comprised of values of variable j measured at time k. After scaling and mean centering of the flattened matrix, a lag of 10 batches is used in matrix reformatting for supervised model building. In other words, we use the prior 10 batches of data to forecast the next batch.

All the time series forecasting models are built using Python. These includes the sequence-to-sequence (seq2seq), support vector machine (SVM) regression, and partial least squares (PLS) models. The libraries used include tensor flow, sklearn, pandas, and numpy.

Table 1 Simulation conditions for batch data generation

Case	Training	Testing
	400 batches with seasonality	200 batches with seasonality

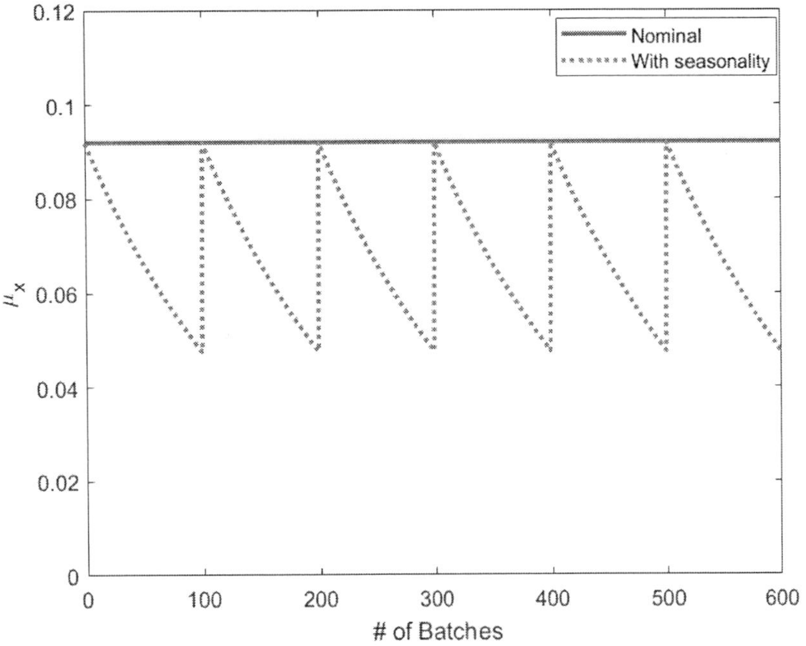

Figure 12 Seasonality in specific growth rate of bacteria culture over six cycles (decay and reset)

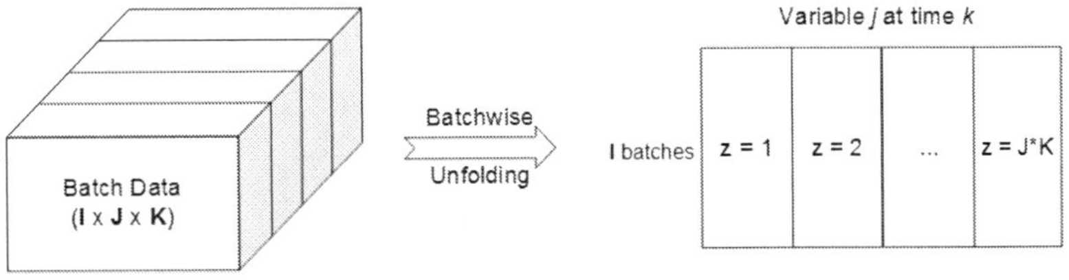

Figure 13 Batch-wise unfolding scheme in time series forecasting

4. Results and Discussions

In this study, we use the root mean square error of forecasting calculated in Equation (8) to evaluate the forecasting performance. The associated results are listed in Table 2. It is observed that the seq2seq model outperforms the other two supervised learning methods in terms of forecast accuracy of batch seasonality.

$$RMSEF = \sqrt{\frac{\sum_{i=1}^{n}(y_i - \hat{y}_i)^2}{n}} \quad (8)$$

Table 2 RMSEF values of forecasting models

Forecasting models	RMSEF
Seq2seq	3.41
SVM	4.40
PLS	4.46

The actual forecasting results are illustrated in Figures 14~16. It is observed that the nonlinear models including the seq2seq and SVM obtain smoother forecasting results than the linear PLS model. In addition, the seq2seq model excels in forecasting the exact batch where the production of penicillin rises abruptly due to reset of the bacterial growth rate.

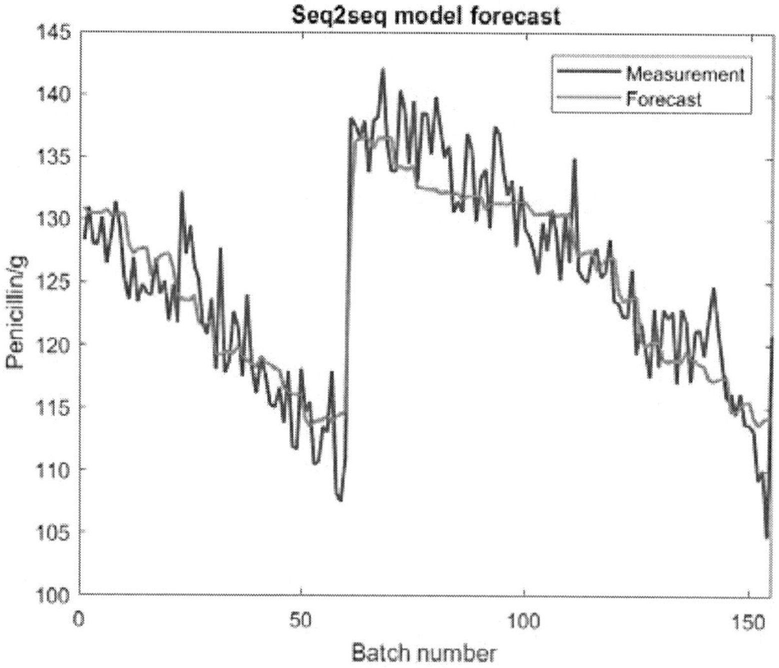

Figure 14 Seq2seq model forecasting results

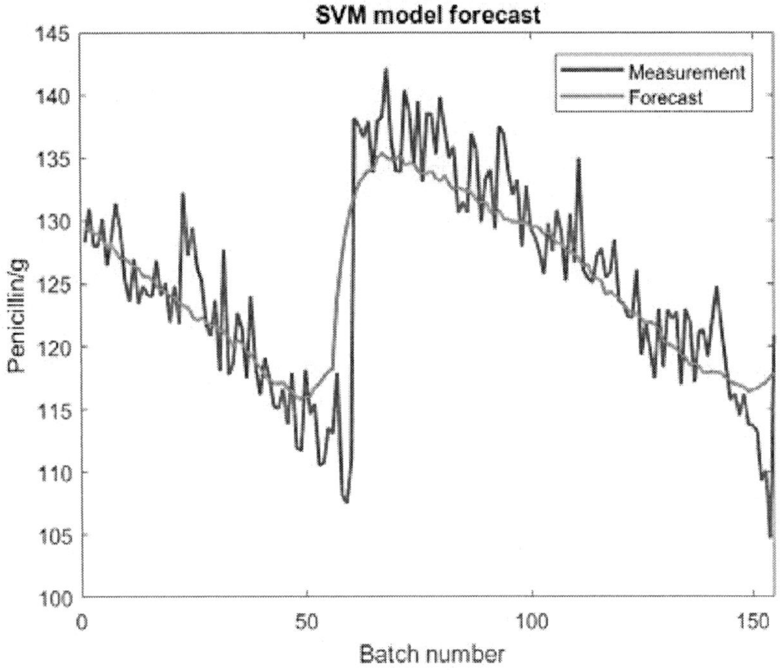

Figure 15 SVM model forecasting results

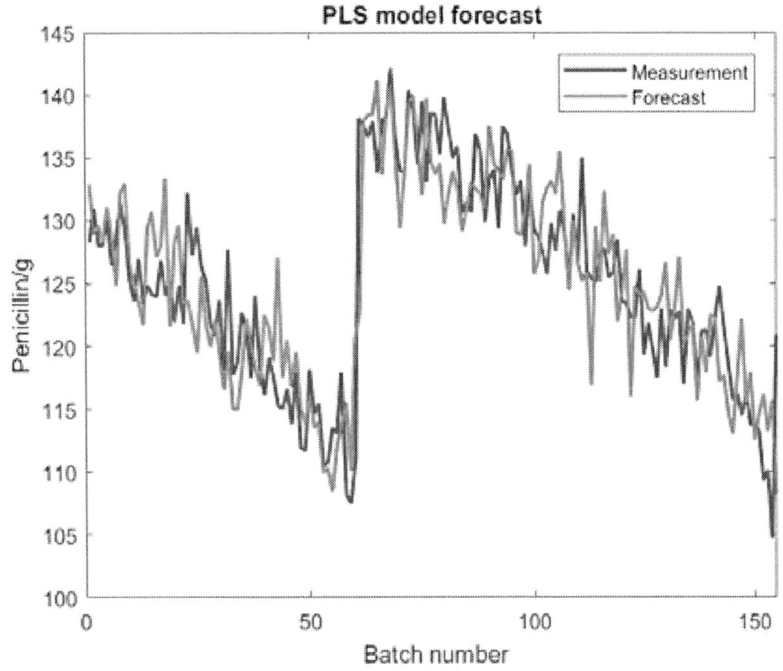

Figure 16 PLS model forecasting results

5. Conclusions and future work

We have demonstrated that the sequence to sequence learning framework can be used to overcome the challenge of batch seasonality forecasting. We built a time series forecasting model in Tensor Flow, and tested it on the data generated from the benchmark penicillin bio-reactor simulator. The model successfully forecast the cyclic change of the penicillin production rate with the periodic decay of the biomass growth rate. Such results illustrate the successful application of data analytics to wrestle with real problems in chemical manufacturing within the digital transformation era.

In the future, we will test such a seq2seq framework on more industrial data and compare its performance with other time series forecasting methods.

References

[1] C. Clarotti, A. Lannoy, S. Odin, and H. Procaccia, "Detection of equipment aging and determination of the efficiency of a corrective measure," *Reliab. Eng. Syst. Saf.*, vol. 84, no. 1, pp. 57–64, Apr. 2004.

[2] M. Ji, Y. He, and T. C. E. Cheng, "Single-machine scheduling with periodic maintenance to minimize makespan," *Comput. Oper. Res.*, vol. 34, no. 6, pp. 1764–1770, Jun. 2007.

[3] B. W. Wojciechowski, "A theoretical treatment of catalyst decay," *Can. J. Chem. Eng.*, vol. 46, no. 1, pp. 48–52, 1968.

[4] H. Müller-Steinhagen, *Heat exchanger fouling: mitigation and cleaning techniques.* IChemE, 2000.

[5] S. J. Qin, "Process Data Analytics in the Era of Big Data," *AIChE J.*, vol. 60, no. 9, pp. 3092–3100, 2014.

[6] G. E. P. Box, G. M. Jenkins, G. C. Reinsel, and G. M. Ljung, *Time series analysis: forecasting and control.* John Wiley & Sons, 2015.

[7] Y. Bengio, P. Simard, P. Frasconi, and others, "Learning long-term dependencies with gradient descent is difficult," *IEEE Trans. neural networks*, vol. 5, no. 2, pp. 157–166, 1994.

[8] S. Hochreiter and J. Schmidhuber, "Long short-term memory," *Neural Comput.*, vol. 9, no. 8, pp. 1735–1780, 1997.

[9] G. Birol, C. Ündey, S. J. Parulekar, and A. Çınar, "A morphologically structured model for penicillin production," *Biotechnol. Bioeng.*, vol. 77, no. 5, pp. 538–552, 2002.

[10] G. Birol, C. Ündey, and A. Cinar, "A modular simulation package for fed-batch fermentation: penicillin production," *Comput. Chem. Eng.*, vol. 26, no. 11, pp. 1553–1565, Nov. 2002.

Appendix – Penicillin Batch Reactor Model

Modeling Equations

The modeling equations and associated parameters and constants are taken from Birol et al. [10] and are simply provided here for completeness and convenience.

Biomass Growth

$$\frac{dX}{dt} = mX - \frac{X}{V}\frac{dV}{dt}$$

where μ is the specific growth rate and contains the effects of environmental variables (pH and Temperature) as well as carbon source (glucose) and oxygen in its kinetic expression.

Penicillin Production:

$$\frac{dP}{dt} = m_{pp}X - KP - \frac{P}{V}\frac{dV}{dt}$$

where μ_{pp} is the specific penicillin production rate containing biomass, carbon source (glucose) and oxygen concentrations in its kinetic expression. Here, the hydrolysis of penicillin is also considered and expressed as a first order rate expression with a rate constant K.

Substrate Utilization:

Glucose:

$$\frac{dS}{dt} = -\frac{m}{Y_{xs}}X - \frac{m_{pp}}{Y_{ps}}X - m_X X + F - \frac{S}{V}\frac{dV}{dt}$$

Oxygen:

$$\frac{dC_L}{dt} = -\frac{m}{Y_{xo}}X - \frac{m_{pp}}{Y_{po}}X - m_o X + K_{la}\left(C_L^* - C_L\right) - \frac{C_L}{V}\frac{dV}{dt}$$

Here, K_{la} is taken to be a function of agitator power input and flow rate of oxygen as suggested by Bailey and Ollis (1986).

Carbon Dioxide Production:

$$\frac{dCO_2}{dt} = a_1\frac{dX}{dt} + a_2 X + a_3$$

The rate expression for CO_2 has been taken from the earlier work of Montague et al. (1986).

Effect of pH

$$\frac{d\left[H^+\right]}{dt} = \gamma\left(\mu X - \frac{FX}{V}\right) + \left[\frac{-B + \sqrt{\left(B^2 + 4\times10^{-14}\right)}}{2} - \left[H^+\right]\right]\frac{1}{\Delta t}$$

Where $B = \frac{\left[10^{-14}/\left[H^{+}\right]-\left[H^{+}\right]\right]V - C_{a/b}\left(F_a + F_b\right)\Delta t}{V + \left(F_a + F_b\right)\Delta t}$, F_a and F_b represents acid and base flow rates in l/h, respectively.

Volume Change

$$\frac{dV}{dt} = F + F_{a/b} - F_{loss}$$

where $F_{a/b}$ is the flow of acid/base, F_{loss} accounts for the evaporative losses during fermentation: $F_{loss} = V\lambda\left(e^{5(T-T_o)/(T_v-T_o)} - 1\right)$

Heat of Reaction

$$\frac{dQ_{rxn}}{dt} = r_{q1}\frac{dX}{dt}V + r_{q2}XV$$

where *r* is the yield coefficient associated with the exothermic microbial reactions.

Energy Balance

$$\frac{dT}{dt} = \frac{F}{s_f}\left(T_f - T\right) + \frac{1}{V\rho c_p} \times \left[Q_{rxn} - \frac{aF_c^{b+1}}{F_c + \left(aF_c^b / 2\rho_c c_{pc}\right)}\right]$$

PID controller

A discretized PID control law has been used in both the temperature and pH controls. E_N denotes the error at time step N and E_{N-1} denotes the error at time step N-1.

$$\Delta MV_N = K_c\left(E_N - E_{N-1} + \frac{\Delta t}{\tau_1}E_N - \frac{\tau_d}{\Delta t} \times \left(CV_N - 2CV_{N-1} + CV_{N-2}\right)\right)MV_N$$

Overall, there are total of 15 differential equations which are solved simultaneously. All the parameters are taken from literature or assigned values which are physically meaningful in the absence of available literature. Because most of the parameters are strain-specific and are also affected by the nature of the substrate and the environmental conditions like pH and temperature, be aware that the parameters used in this simulator are average values rather than being specific to a certain strain or an environmental condition.

Initial Conditions, Kinetic and Controller Parameters

Initial conditions	
Substrate concentration: S (g/l)	15
Dissolved oxygen concentration: C_L ($= C_L^*$ at saturation) (g/l)	1.16
Biomass concentration: X (g/l)	0.1
Penicillin concentration: P (g/l)	0
Culture volume: V (l)	100
Carbon dioxide concentration: CO_2 (mmol/l)	0.5
Hydrogen ion concentration: $[H^+]$ (mol/l)	$10^{-5.1}$
Temperature: T (K)	297
Heat generation: Q_{rxn} (cal)	0
Kinetic parameters and variables	
Feed substrate concentration: s_f (g/l)	600
Feed flow rate of substrate: F (l/h)	
Feed temperature of substrate: T_f (K)	298
Yield constant: $Y_{x/s}$ (g biomass/g glucose)	0.45
Yield constant: $Y_{x/o}$ (g biomass/g oxygen)	0.04
Yield constant: $Y_{p/s}$ (g penicillin/g glucose)	0.90
Yield constant: $Y_{p/o}$ (g penicillin/g oxygen)	0.20
Constant: K_1 (mol/l)	10^{-10}
Constant: K_2 (mol/l)	7×10^{-5}
Maintenance coefficient on substrate: m_x (per h)	0.014
Maintenance coefficient on oxygen: m_o (per h)	0.467
Constant relating CO_2 to growth: α_1 (mmol CO_2/g biomass)	0.143
Constant relating CO_2 to maintenance energy: α_2 (mmol CO_2/g biomass h)	4×10^{-7}
Constant relating CO_2 to penicillin production α_3 (mmol CO_2/l h)	10^{-4}
Maximum specific growth rate: μ_x (per h)	0.092
Contois saturation constant: K_x (g/l)	0.15
Oxygen limitation constant: K_{ox}, K_{op} (no limitation)	0
Oxygen limitation constant: K_{ox}, K_{op} (with limitation)	2×10^{-2}, 5×10^{-4}
Specific rate of penicillin production: μ_p (per h)	0.005
Inhibition constant: K_p (g/l)	0.0002
Inhibition constant for product formation: K_I (g/l)	0.10
Constant: p	3
Penicillin hydrolysis rate constant: K (per h)	0.04
Arrhenius constant for growth: k_g	7×10^3
Activation energy for growth: E_g (cal/mol)	5100
Arrhenius constant for cell death: k_d	10^{33}
Activation energy for cell death: E_d (cal/mol)	50 000
Density × heat capacity of medium: $\rho\ C_p$ (per l °C)	1/1500
Density × heat capacity of cooling liquid: $\rho_c\ C_{pc}$ (per l °C)	1/2000
Yield of heat generation: r_{q_1} (cal/g biomass)	60
Constant in heat generation: r_{q_2} (cal/g biomass h)	1.6783×10^{-4}
Heat transfer coefficient of cooling/heating liquid: a (cal/h °C)	1000
Cooling water flow rate: F_c (l/h)	
Constant: b	0.60
Constants in K_{la}: α, β	70, 0.4
Constant in F_{loss}: λ (per h)	2.5×10^{-4}
Proportionality constant: γ (mol $[H^+]$/g biomass)	10^{-5}

Controller parameters (PID)	
pH: (base) K_c, τ_I: (h), τ_d: (h)	8×10^{-4}, 4.2, 0.2625
pH: (acid) K_c, τ_I: (h), τ_d: (h)	1×10^{-4}, 8.4, 0.125
Temperature: (cooling) K_c, τ_I: (h), τ_d: (h)	70, 0.5, 1.6
Temperature: (heating) K_c, τ_I: (h), τ_d: (h)	5, 0.8, 0.05

554324 Text Mining and Analytics (poster)

Text mining seeks to extract information and insights from textual data. In particular the distillation and summary of documents is of increasing commercial interest, from news media to patents to maintenance logs.

Application areas include business analysis, scientific discovery, predictive maintenance, marketing and communications, litigation, and cybersecurity, among others.

Basic workflow elements will be reviewed in this talk, including: definition of the corpus (which may involve the pre-step of converting images to text), preprocessing, visualization, filtering, statistical analysis, word embeddings, and classification.

Some more advanced subjects, for example topic modeling and sentiment analysis, will be introduced and described.

Demonstrations of functionality will be presented using MATLAB.

551750 Applying Optimization Method to Deliver Proactive Pricing on Improving Margin Beyond a Cost Plus Strategy

The Dow Chemical Company is a leading chemical manufacturer in the world. Dow manufactures plastics, chemicals, and agriculture products. With a presence in over 150 countries, it employs over 50,000 people worldwide. Price is a function of volume, cost, previous price and external factors. We have provided a dynamic advanced analytical engine that can be coupled with a Product Manager's current process to maximize margin potential. This tool utilizes Price-volume elasticity while including capacity constraints and necessary external variables defined by the BU product organization.

550013 An Holistic Approach for Batch Analysis and Monitoring (Poster)

ABB Ability^{TM} BatchInsight identifies and monitors anomalies in the batch process. It helps process specialists and plant operators detect and resolve abnormalities by conducting batch-level analysis.

Using time-level analysis, it recognizes the time of occurrence of the deviation and isolates the process variables that contributed to the deviation, enabling effective and timely corrective action.

The big data algorithm empowered batch monitor supervises continuously the running batch process. It enriches a normal PV monitor with additional features like, similarity-based process forecasts, adaptive "golden batch" process corridor, batch-phase change recognition & visualization, and operator decision support after early warning.

549714 Best Big Data Analytics Platform for Manufacturers (Poster)

Artificial Intelligence, Machine Learning, and the Industrial Internet of Things are still the hot topics in the big data analytics world. Manufacturers have been generating data off the line for decades, so how can manufacturers get the most of these new advancements in analytics? The combination and convergence of new hardware infrastructures have allowed for the advancement of analytics like artificial intelligence to move from research out into the business world. Analytic insights need to be deployed into production and actions need to be taken sometimes in real-time to truly realize the benefits of all these advancements. Come see how the power of the SAS platform with it's support of the entire analytics lifecycle: Data, Discovery, and Deployment provides the best big data analytics platform for manufacturers to leverage all sorts of analytics both in batch and in real-time.

551864 Microsoft - Empowering the Chemical Industry (Poster)

Chemical companies are revisiting their own business processes and technology approaches as competitors and partners start to employ 'digitalized' business processes and exploit the increasing convergence between OT and IT on the plant floor, to connect the enterprise as a whole to the extended supply chain and throughout the ecosystem.

In this session you will learn how technology empowers workforce with devices, while AI-powered productivity tools enable them to work more effectively, manage manufacturing complexity, and collaborate specifically to their business and operational needs. Understand Microsoft approach to Firstline Worker including Mixed Reality with introduction to Microsoft Remote Assist and Layout.

551860 Business Value, Work Process, Analytics and Data in the Chemical Industry (Poster)

This presentation will begin with a discussion of the major categories of business value that can be unlocked by applying analytics to production activities, including asset reliability, production optimization, energy efficiency optimization and safety performance improvement. It will then explore how value is realized through transformation of work processes to identify several key requirements of digitalization programs and analytics, suggesting answers questions such as:

- Should I use the cloud for analytics?
- Is my process and predictive technology data sufficient to support analytics?
- Do I need to use artificial intelligence?
- How important are metadata (metadata are plural!)?

The presentation will include a demonstration of first-principles-based analytics approach applied to early detection of asset failure in support of a top quartile maintenance work management process.

551856 Prognostic and Health Management Using Digital Twin (Poster)

Digital Twin is a digital representation of your real-world device or operation that mirrors actual equipment and operating conditions. Leveraging this technology for your equipment's predictive health and operational optimization can significantly improve your company's EBITDA.

However, creating digital twins for your assets continues to be a costly affair. A typical refinery or plant has hundreds of assets. Modeling them in a robust manner is a herculean task. We will present a methodology driven by AI that works with variations in model types within a class of assets but also across different types of assets. As a result, you can rapidly scale analytics across your organization delivering extraordinary savings to your company's bottom line. Our patent pending Asset Agnostic Prognostic Technology has shown the efficiency of this approach across a variety of mechanical equipment (pumps, compressors and motor bearings. DeepIQ's prognostic health management software is

1. *a) Asset Agnostic: is easily configurable for new assets*
2. *b) Source Agnostic: connects to a wide variety of time series industrial sources*
3. *c) Platform Agnostic: works natively on both on-premise or Cloud big data environments.*

550567 Use of Process Imaging to Improve the Performance of a Filter

Solid-liquid separation is rate-limiting step for many DOE waste treatment processes. Understanding the fouling structure and its relationship to the foulant may shed light into improving cleaning cycles (minimizing waste streams) and optimizing filter performance (extended lifetime operations). The work used process imaging technologies to quantify dynamics of solid particle movement and filter cake formation on surface of filters, and measure the shape, thickness, dimensions, and buildup of the filter cake as a function of time.

551199 Effect of Different Melt Mixing Methods on the Thermomechanical and Rheological Properties of LLDPE/HNT Nanocomposite

In polymer nanocomposite production good mixing is always desired to attain good material properties. Beside the interactive chemistry of constituents in the mixture, the condition of interface between components and the degree of dispersion strongly influence the performance of final nanocomposite.

Polymeric nanocomposites prepared with halloysite nanotube (HNT) nanofillers have gained strong interest in recent years. HNT is a natural nanotube which is abundantly available at a low price. Compared to other nanoparticles, it has better dispersion due to less inter-tubular contact and its unique geometry.

This experimental research focuses on fabricating nanocomposite with linear low density polyethylene (LLDPE) and HNT with different amounts of HNT loading and using different melt mixing methods. The melt mixing method among all other available techniques is the most compatible with the current industrial compounding and processing technology which has environmentally friendly nature. In this study the effects of different melt mixing methods on the thermomechanical and rheological properties of prepared LLDPE/HNT nanocomposites will be elucidated.

551157 The Effect of Temperature on Flow Field in a Cyclone Separator By Numerical Simulation

The effect of temperature on flow field in a cyclone separator by numerical simulation

Qing Wei[a,b]; Guogang Sun[a,b,*]; Yunnan Jiao[a,b]

[a]*State Key Laboratory of Heavy Oil Processing, China University of Petroleum, Beijing 102249, China*

[b]*Beijing Key Laboratory of Process Fluid Filtration and Separation, Beijing 102249, China*

Corresponding author: Guogang Sun; Tel: +86-10- 89734820;

E-mail:ggsunbj@163.com

Abstract: Temperature is one of the most important parameters for the performance of a cyclone separator. In this study, a computational fluid dynamics is used to predict the flow field in a volute cyclone separator using the Reynolds stress transport model when temperature ranges from 293 K from 1223 K. The numerical results are compared with the experimental data and good agreement is observed. The results show as temperature increases, the tangential velocity and radial velocity of cylindrical and the conical sections decrease, but axial velocity remains almost to be the same. The temperature increase results in a decrease of the tangential velocity along the axial height, and the amplitude of velocity fluctuation. Meanwhile the inner vortex size is reduced at high temperatures. With temperature increasing, natural vortex length and the frequency of vortex end first decreases dramatically and then remains essentially unchanged after temperature above 823K. The effect of temperature on gas flow is achieved by changing the gas viscosity and density. As temperature increases, Reynolds number decreases with the inlet velocity and dimensions of cyclone separator unchanged, which is the root cause of the flow field changing.

Key words: cyclone separator; temperature; flow field

References

- Jianyi Chen, Chunxi Lu, Mingxian Shi. Experimental measurement on flow field in a high-temperature cyclone separator. 2010, 61(9): 2340-3245.
- P A Patterson, R J Munz. Gas and particle flow patterns in cyclones at room and elevated temperatures. Canadian Journal of Chemical Engineering, 2010, 74(2): 213-221.
- M S Shi, H S Kim, D S Jang. A numerical and experimental study on a high efficiency cyclone dust separator for high temperature and pressurized environments. Appl. Therm. Eng. 2006, 25(12): 1821-1835.
- Parker R, Jain R, Calvert S. Particle collection in cyclones at high temperature and high pressure. Environ. Sci. Technol., 1981, 15(4): 451-458.

546646 Dynamic Performance of Flow Field in an Cylindrical Cyclone Separator

This study presents a flow analysis on an atypical cylindrical cyclone separator with a large ratio of length to diameter. Due to the distribution of the flow field determines the merits of separation performance, the aim of this study is to optimize the flow. The Phase Doppler Particle Analyzer (PDPA) is used to measure the gas flow, and the numerical simulation is carried out by using the Reynolds Average Navier-Stokes (RANS) equation, the Reynolds stress model (RSM). Since the model is validated by good agreement between the numerical results and experimental data, and then the tangential velocity is analyzed from the transient flow field and dynamic performance.

The results showed that the tangential velocity of the transient flow field existed a non-axisymmetric phenomenon in the cross section, which was mainly presented as asymmetric distribution of the contours. The zero location of tangential velocity value was not coincide with the geometric center, where the tangential velocity was larger near the zero location. In addition, the dynamic character of tangential velocity was high-speed pulsating, with quasi-periodic characteristic. The structure or operation optimization could change the pulsation frequency and reduce the industrial vibration, which is highly significant.

Meanwhile, a parameter of RMS was used to identify the degree of turbulence velocity deviating from mean velocity. The RMS data showed that the flow instability gradually increased with the fluid flowing into the cyclone separator from the inlet section, until reached a peak value at the position about 2.0 times the length to diameter. And then the flow instability gradually decreased, with the fluid energy dissipated.

551222 Reactor Performance Assessment Using Experimental Studies and Multiscale Modeling: Liquid Phase Ethylene Production By Hydrogenation of Acetylene

Trickle bed and two-phase upflow packed bed reactors have found extensive applications in petroleum, petrochemical, chemical and environmental industries especially for processes involving complex mechanisms. Assessing the reactor performance for these reactors needs extensive experimentation complemented by validation by a reliable mathematical model at different scales. Various factors like incomplete catalyst wetting, liquid maldistribution, interphase mass transfer, intraparticle diffusion are necessary to predict the reactor performance. Moreover, accounting the variation in the overall effectiveness factor across the reactor axis due to the change in the local reactant concentrations is also significant. A one-dimensional mathematical axial dispersion flow model integrated with a pellet scale model has been developed. The kinetics and the related parameters involved in the processes were accounted in the model along with different particle to liquid wetting conditions and correlations. The model was validated using different experimental studies at a range of operating conditions from literature. The hydrogenation of acetylene in the liquid phase for ethylene production will be considered as the model reaction to assess the reactor performance and help in obtaining meaningful scaleup parameters. Gas phase catalytic hydrogenation of acetylene to produce ethylene is commonly practiced in industries to increase the ethylene yield from the cracker. This process has green oil formation which leads to catalyst deactivation and sometimes reactor runaway risks. To overcome this issue as well as to increase the selectivity and conversion liquid phase hydrogenation is under current research. The catalyst life and stability increases by reducing the green oil formation. The solvent (liquid) selectively absorbs acetylene from the cracked effluent, and this liquid phase improves heat transfer in the reactor. The kinetics studies were investigated in a stirred tank reactor in both slurry and basket modes with liquid in batch and gas as continuous using a Pd/Al_2O_3 catalyst . N-Methyl Pyrrolidone (NMP), a polar solvent, was used as the liquid phase. The reaction was conducted at a pressure of 250 psig and temperatures ranging between 60-120^oC. In order to estimate the kinetic parameters, a model was developed based on Langmuir-Hinshelwood-Hougen-Watson approach. The dispersion parameter and liquid holdup obtained from the residence time distribution studies by conducting cold flow experiments using a liquid tracer will be integrated into the model. At Multiphase Flows and Reactors Engineering and Applications laboratory (mFReal), a high-pressure, high temperature unit has been built to conduct sensitive reaction experiments. This unit is equipped with a blast shield and has laboratory scale reactors with thermowell and gas-liquid distributors. The unit will be tested for selected operating conditions for the acetylene hydrogenation in the liquid phase to demonstrate the safe experimentation as well as improving the conversion of the process. The experimental data will be validated using the developed multiscale model to assess the reactor performance and further optimize the operating parameters, process time and number of trials.

Application of GRAND and 2DGR Technique Method in Risk Assessment of Hydrogen Peroxide Production Process

Siyu Sui
School of Mechanical and Electrical Engineering
China University of Petroleum（East China）, Qingdao, China
siyuaaane@163.com

Zhicheng Chu
School of Mechanical and Electrical Engineering
China University of Petroleum（East China）, Qingdao, China

Yufeng Kan
School of Mechanical and Electrical Engineering
China University of Petroleum（East China）, Qingdao, China

Dongfeng Zhao
School of Chemical Engineering
China University of Petroleum（East China）, Qingdao, China

Prepared for Presentation at
American Institute of Chemical Engineers
2019 Spring Meeting and 15th Global Congress on Process Safety
New Orleans, LA
March 31 – April 3, 2019

Application of GRAND and 2DGR Technique Method in Risk Assessment of Hydrogen Peroxide Production Process

Siyu Sui
School of Mechanical and Electrical Engineering
China University of Petroleum（East China）, Qingdao, China
siyuaaane@163.com

Zhicheng Chu
School of Mechanical and Electrical Engineering
China University of Petroleum（East China）, Qingdao, China

Yufeng Kan
School of Mechanical and Electrical Engineering
China University of Petroleum（East China）, Qingdao, China

Dongfeng Zhao
School of Chemical Engineering
China University of Petroleum（East China）, Qingdao, China

Keywords: hydrogen peroxide, inherent safety, Graphical Descriptive Technique for Inherent Safety Assessment (GRAND), 2-Dimensional Graphic Rating (2DGR), risk assessment

Abstract

Hazard reduction and elimination play a major role in designing a user-friendly plant in order to prevent accidents. To achieve this, risk assessment on the properties of the chemicals and the process during the early design stage is important to understand the hazards posed by the process involved. As we all know, Hydrogen peroxide, as a green environmentally friendly oxidant, is widely used in various fields of industry, and it causes a series of serious fire and explosion accidents due to its thermal decomposition explosion risk. In this paper, based on the principle of inherent safety, Graphical Descriptive Technique (GRAND) and the 2-Dimensional Graphic Rating technique (2DGR) technique method were applied on three processing routes for Hydrogen Peroxide production as a case study. Moreover, the root-cause of hazards were further analyzed. Among them, through the essential risk assessment of the three production processes, the hazard levels of different processes were identified and quantified. The 2DGR technology was used to analyze the two hydrogen peroxide production processes, such as the enthalpy method with similar process hazard level and the direct hydrogen synthesis method. The hydrazine method is the most dangerous process route. The root-cause of hazards analysis of the production process of hydrazine was used to determine the flammability and explosiveness of hydrogen. The short-term toxicity contact threshold of hydrogen peroxide and heavy aromatic hydrocarbons and the reactivity of hydrogen peroxide were the root causes of the hazard of the process route. Not only can quantitative risk assessment of hydrogen peroxide process, but also risk root cause analysis, which has important theoretical value for the inherently safer design and production of hydrogen peroxide.

1 Introduction

Hydrogen peroxide, as an oxidant, will excite its heat hazard under high temperature or with some incompatible chemicals, which will lead to thermal runaway reactions, which will eventually lead to explosion accidents[1]. Therefore, the process route of hydrogen peroxide is carried out. Risk assessment has important practical significance. Risk assessment methods can be divided into qualitative, semi-quantitative and quantitative risk assessment. Commonly used risk assessment methods include HAZOP method, Failure Mode and Effect Analysis(FMEA), Fault Tree Analysis(FTA), Dow Chemical Fire and Explosion Hazard Index method etc..

Li Qi et al[2] analyzed the hydrogen peroxide device using Hazard and Operational Analysis (HAZOP), and identified the hidden dangers of the device and proposed corresponding measures. This approach establishes the idea of a hydrogen peroxide system rather than a single device. However, HAZOP itself has the disadvantages of time consuming, laborious, high cost, and excessive reliance on experience. Mao Gang [3] used the fire and explosion risk index evaluation method to conduct a risk assessment of the hydrogen peroxide production device, identify the risk of the device, and truly quantify the expected loss of potential fire, explosion and reactive accidents, and determine the possible accident. The device that has occurred or the accident has been expanded, and the improvement measures have been proposed to eliminate the safety hazard; this method has the disadvantage of subjective quantification of the value of the hazard coefficient and the compensation coefficient, and the management factor is also considered less, so it is mainly for the chemical production process. Ebrahimi [4] et al. not only used the different production processes for the production of hydrogen peroxide by the method of hydrazine, but also the two production processes for the production of hydrogen peroxide by the direct method of hydrazine and hydrogen and oxygen, using a strengthening process for safety comparison. Based on the concept of the safety protection layer used by the inventory tool to detect the positive and negative effects after the reinforcement process, and intuitively determine the safety situation before and after the process is strengthened, but there is no specific quantitative indicator, but Ahmad [5] et al. Based on the safety method, GRAND [6] is proposed for the inherent safety assessment of the process. The physical property data of the chemical substance used and the operating conditions of the production process are quantitatively evaluated by generating a quantitative score by a mathematical model. This approach solves the problem of subjective quantification of many evaluation methods. According to the parameter data of the evaluation system, the evaluation results can be reproducible, and the problem that the data cannot be reproduced when the experts rely on experience for risk evaluation is avoided. In this paper, the GRAND risk assessment method is used to evaluate the inherent safety of different hydrogen peroxide production methods, identify and quantify the hazard sources in the process, and define the risk levels of different production processes.

2 Mathematical model

Aside from root-cause analysis, GRAND is extended to 2DGR to represents different level of hazards to the process routes evaluated. Ranking of process routes solely based on the final assessment scores is not sufficient in differentiating the level of hazards as each route might only have a few scores difference to one another.

Through the GRAND mathematical model, the scores of each parameter of each process route are obtained, and then the hazard sources of different production routes are further evaluated. The GRAND method divides safety into two parts, chemical safety and process safety[7]. The four parameters considered for chemical safety are flammability, explosiveness, toxicity, and reactivity. Process safety parameters also consists of four parameters which are the temperature, pressure, heat of reaction and process inventory parameters.

2.1 Chemical safety parameter

2.1.1 Flammability parameter

Liquids with lower flash points are more hazardous than liquids with higher flash points. The logistic equation for flammability is produced as Eq.(1). S_{FL} refers to the scores for the flammability parameter while X_{FL} refers to the flash point value to be evaluated. Fig.1(1) illustrates the flammability curve which indicates that lower flash point temperature will results in higher score.

$$S_{FL} = 100 \times \left(1 - \left(\frac{1}{1 + 3.03e^{-0.02X_{FL}}}\right)\right)$$

——Eq.(1)

2.1.2 Explosiveness parameter

Wider range between LEL and UEL indicates higher tendency for explosion which repre-sents by logistic equation in Eq. (2). In Eq. (2),X_{EXP} is the explosiveness value (or in GRAND the difference between UEL and LEL), while S_{EXP} is the score produced for the explosiveness value evaluated. Fig.1(2) is plotted with higher range between UEL and LEL, which indicates higher hazard with high score designation.

$$S_{EXP} = 100 \times \left(\frac{1}{1 + 1096.63e^{-0.14X_{EXP}}}\right)$$

——Eq.(2)

2.1.3 Toxicity parameter

One indicator that can be used in determining the toxicity of a chemical established by the American Conference of Governmental Industrial Hygienists (ACGIH) is called threshold limit values (TLVs). In this method, threshold limit values for short-term exposure limit (TLV-STEL) are used, which is more significant for an acute toxicity type of event. Since a chemical with a lower TLV-STEL value is more hazardous than a chemical with a higher TLV-STEL value, the equation needs to be subtracted from 1 before the whole equation is multiplied by coefficient C, which is 100. Eq. (3) shows the logistic equation for the toxicity parameter. The score produced is denoted as S_{TOX}, while X_{TOX} de-notes the TLV-STEL value to be evaluated. Fig.1(3) shows the GRAND curve plotted with higher TLV-STEL values, represented by lower score values.

$$S_{TOX} = 100 \times \left(1 - \left(\frac{1}{1 + 403.4288e^{-0.012X_{TOX}}}\right)\right)$$

———Eq.(3)

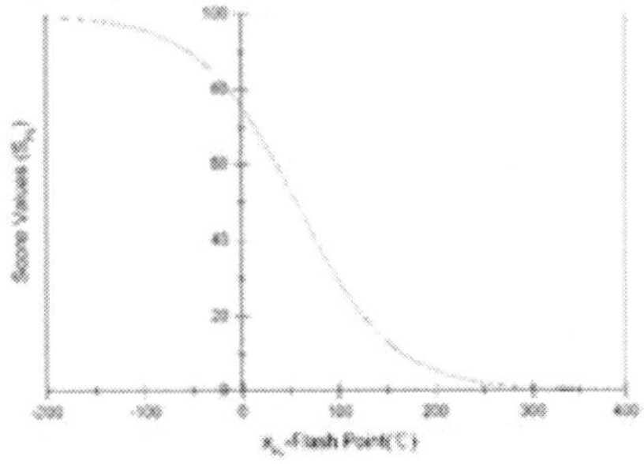

(1) GRAND Curve for Flammability Parameter

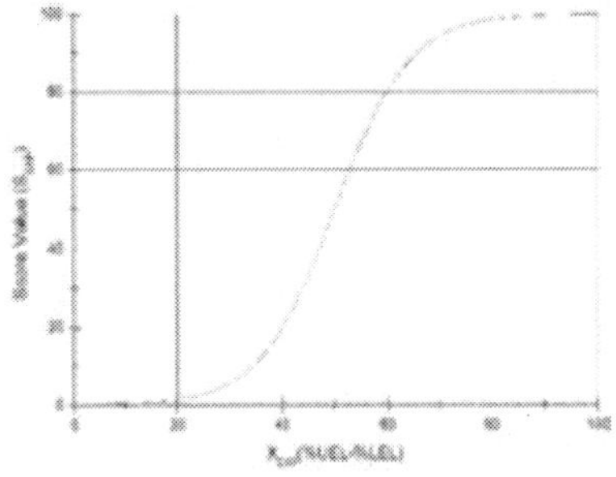

(2) GRAND Curve for Explosiveness Parameter

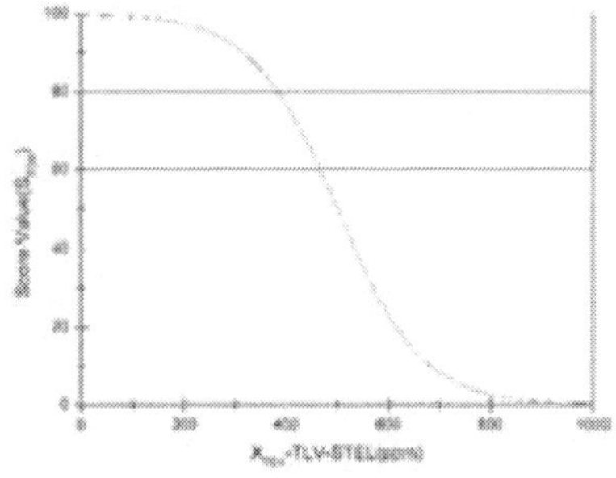

(3) GRAND Curve for Toxicity Parameter

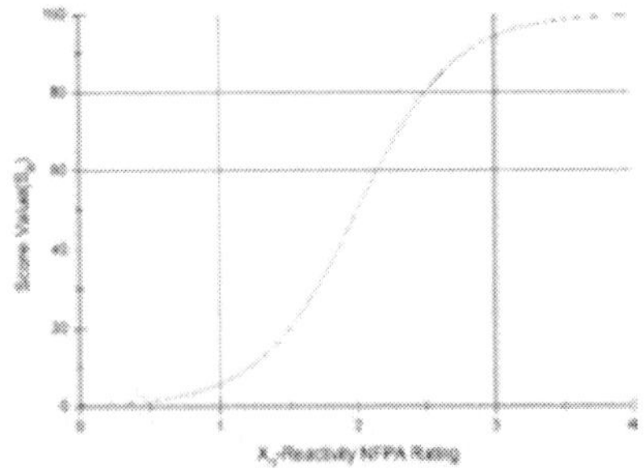

(4) GRAND Curve for Reactivity Parameter

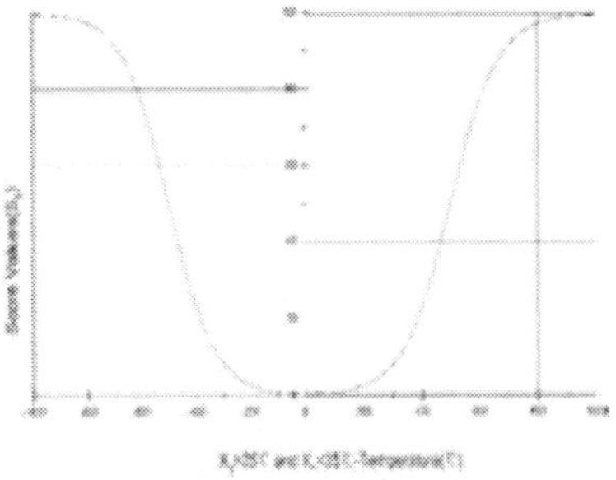

(5) GRAND Curve for Temperature Parameter

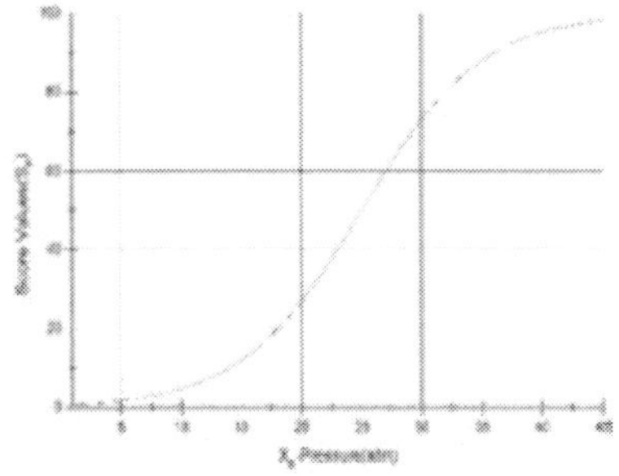

(6) GRAND Curve for Pressure Parameter

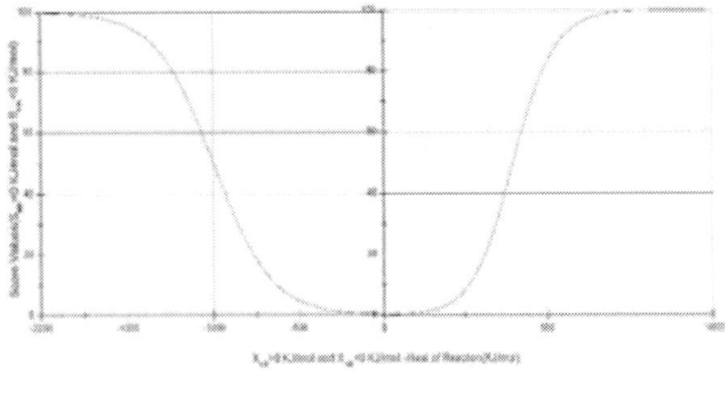

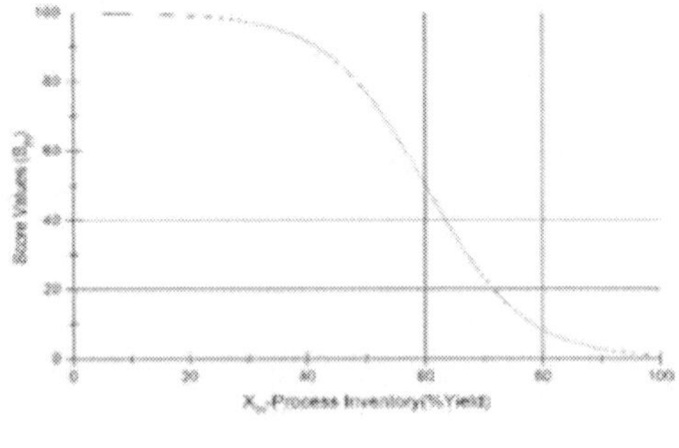

(7) GRAND Curve for Heat of Reaction Parameter

(8) GRAND Curve for Process Inventory Parameter

Fig. 1. Graphical and numerical descriptive (GRAND) curve.

2.1.4 Reactivity parameter

NFPA reactivity rating is used to measure reactivity. NFPA rating is widely available and is derived from the Material Safety Data Sheet (MSDS) that come with the chemicals. The hazard rating index for reactivity is summarized in Table 1. The logistic equation for reactivity GRAND is as shown in Eq.(4); Fig.1(4) shows the plotted curve for the equation. X_R in Eq. (4) represents the NFPA reactivity rating while SR represents the score produced for the NFPA reactivity rating.

$$S_R = 100 \times \left(\frac{1}{1 + 270.43e^{-2.8X_R}}\right)$$

———Eq.(4)

Table 1

Hazard rating index: Reactivity [9].

Rating Indication Explanation		
0	Stable	Not reactive when mixed with water.
1	Caution	May react if heated or mixed with water but not violently.
2	Warning	Unstable or may react violently if mixed with water.
3	Danger	May be explosive if shocked, heated under confinement or mixed with water.
4	Danger	Materials which in themselves are readily capable of detonation or of explosive decomposition or explosive reaction at normal temperature and pressures. This degree should include materials which are sensitive to mechanical or localized thermal shock at normal temperatures and pressures.

2.2 *Process safety parameter*

2.2.1 Temperature parameter

Two GRAND curves were developed, GRAND for operating temperature higher than 25℃ and lower than 25℃ with one assumption that the process will not contribute to hazards when operating at the ambient temperature of 25℃. Thus, by taking this assumption, the temperature values gathered need to be subtracted with the value of 25℃ before calculating the scores as shown in Table 2. Eq.(5) and (6) represents the logistic equation for operating temperatures higher and lower than 25℃. S_T>25℃and S_T<25℃ are the score produced for X_T>25℃ and X_T

<25℃ evaluated for operating temperatures higher and lower than 25℃, respectively. Fig.1(5) shows higher scores, indicating higher hazards to the process.

$$S_{T>25°C} = 100 \times \left(\frac{1}{1 + 403.43e^{-0.012X_{T>25°C}}}\right)$$

——Eq.(5)

$$S_{T<25°C} = 100 \times \left(1 - \left(\frac{1}{1 + 0.0025e^{-0.012X_{T<25°C}}}\right)\right)$$

——Eq.(6)

2.2.2 Pressure parameter

A process with higher operating pressure is more hazardous compared to a process with low operating pressure. Eq.(7) shows the logistic equation for the pressure parameter. Evaluated operating pressure, X_P will result in a GRAND score for the pressure parameter, S_P. In Fig.1(6), a high pressure value is represented by a high score value, indicating a greater hazard.

$$S_P = 100 \times \left(\frac{1}{1 + 148.41e^{-0.2X_P}}\right)$$

——Eq.(7)

2.2.3 Heat of reaction parameter

As illustrated in Fig.1(7), the score is higher when the heat of reaction value moves farther away from zero. Eq. (8) and (9) show the logistic equations of the heat of reaction parameter, while Fig. 1(g) shows the curves of the heat of reaction parameter. The X_{HR} symbols in both Eq. (8) and (9) denote the evaluated value for the heat of reaction parameter for higher and lower than 0 kJ/mol while the S_{HR} symbols denotes the scores of the evaluated heat of reaction.

$$S_{HR>0kj/mol} = 100 \times \left(\frac{1}{1 + 601.85e^{-0.016X_{HR>0kj/mol}}}\right)$$

——Eq.(8)

$$S_{HR<0kj/mol} = 100 \times \left(\frac{1}{1 + 403.43e^{-0.006X_{HR>0kj/mol}}}\right)$$

——Eq.(9)

2.2.4 Process inventory parameter

High percentage yield indicates a higher amount of the desired product formed, and thus less of the raw material inventory is needed[8]. The logistic equation for the process inventory parameter is as written in Eq. (10), where X_{PI} refers to the percentage yield value while S_{PI} refers to the score for the evaluated X_{PI}. Fig.1(8) indicates that lower percentage yield will result in a higher score.

$$S_{PI} = 100 \times \left(1 - \left(\frac{1}{1 + 1339.43e^{-0.12X_{PI}}}\right)\right)$$

———Eq.(10)

Using logistic function, S curve profile is generated for all parameters. However, the differences can be seen in the logistic equations produced for each parameters as well as the range of x-axis in Fig. 1.

In this study, GRAND Total Score represents the final scores while GRAND Ranking Curve represents ranking. The GRAND Total Score is the summation of scores for all parameters. This score will be used to represent the routes assessed for ranking purposes, using the GRAND Ranking Curve.

Chemical Safety Total Score (CSTS) is derived using Eq. (11) based on the worst-case scenario. The maximum score received by chemicals for flammability (S_{FL}), explosiveness (S_{EXP}), toxicity (S_{TOX}) and reactivity (S_R) are summed up to represent the reaction step for those particular routes.

$$\text{CSTS} = (S_{FL})_{max} + (S_{EXP})_{max} + (S_{TOX})_{max} + (S_R)_{max}$$

———Eq.(11)

Process Safety Total Score (PSTS) identified for temperature (ST), pressure (SP), heat of reaction (SHR) and process inventory (SPI) parameters are calculated according to Eq. (12). If there are several reaction steps, the mod scores from all reaction steps in a route are taken into account.

$$\text{PSTS} = (S_T)_{max} + (S_P)_{max} + (S_{HR})_{max} + (S_{PI})_{max}$$

———Eq.(12)

GRAND Total Score can be calculated by summing up both total scores as shown in Eq. (13).

$$\text{GRAND Total Score} = \text{CSTS} + \text{PSTS}$$

———Eq.(13)

3 Case study

3.1 Process risk assessment

At present, there are three main routes for the production of hydrogen peroxide in China: the hydrazine method, the isopropanol method, and the direct hydrogen-oxygen synthesis method. In the hydrogenation process, 2-alkyl hydrazine is used as a support, palladium (or nickel) is used as a catalyst, and 2-alkyl hydrazine is reduced to 2-alkylhydroquinone by hydrogen, followed by an oxidation process. In the hydrogenation process, 2-alkyl hydrazine is used as a support, palladium (or nickel) is used as a catalyst, and 2-alkyl hydrazine is reduced to 2-alkylhydroquinone by hydrogen, followed by an oxidation process. 2-alkyl anthraquinones hydrogen is oxidized oxygen in the air or oxygen into the original 2-alkyl anthraquinone while forming hydrogen peroxide.

The isopropanol method uses isopropanol as a raw material, hydrogen peroxide or other peroxide as an initiator, and liquid phase oxidation with air or oxygen to form H_2O_2 and acetone; direct synthesis of hydrogen and oxygen is an acidic solution in which a catalyst is suspended. Hydrogen and oxygen are introduced into the reactor to directly synthesize H_2O_2 by reacting hydrogen and oxygen under the catalyst.

First, in order to explain the risk assessment method, hydrogen peroxide is produced by the hydrazine method, and three main steps of hydrogenation, oxidation, and extraction are quantified. Next, the physical property data and operating conditions of the chemicals in each of the three steps are collected, and the data of each parameter is brought into the eq. (1) to (10) to obtain a GRAND score. The GRAND score for chemical safety and process parameter safety is shown in Tables 2 and 3. The highest score of chemical or reaction conditions for each chemical safety parameter and process safety parameter at all reaction steps represents the highest level of hazard in the process route. For example, in the flammability parameter, the fraction of hydrogen is 96, which is the highest compared to the fraction of matter in all reaction steps. Therefore, 96 represents the highest risk score for the flammability parameter in the process route, which can also be called $(S_{FL})_{max}$. The other seven parameters are used in the same way as the above method. According to $(S_{FL})_{max}$, $(S_{EXP})_{max}$, $(S_{TOX})_{max}$, $(S_R)_{max}$, $(S_T)_{max}$, $(S_P)_{max}$, $(S_{HR})_{max}$ and $(S_{PI})_{max}$ can be used to determine the CSTS and PSTS of eq. (11) and (12). According to Tables 2 and 3, the CSTS in the hydrogen peroxide production route based on the hydrazine method is 296.17, the PSTS is 78.90, and the GRAND score is 375.07. This calculation method is reused in other process routes to obtain relevant data.

The chemical process safety total scores, process safety total scores and GRAND total scores of the three hydrogen peroxide process routes are shown in Table 4. The higher the GRAND total score indicates that the process route is more dangerous, and the lower the score, the safer the process route. R2 is determined to be the safest route with the lowest GRAND total score of 211.86. The GRAND total score of R1 is 375.07, the GRAND total score of R3 is 364.90, and the GRAND total score of R1 is 10.17 more than R3. It can be seen that the dangerous levels of the two processes may be similar. This issue can be resolved by the GRAND curve ranking. The GRAND ranking curve[5] is optimized with existing scores to obtain a more reasonable level of danger. The GRAND curve of different hydrogen peroxide production routes ranks R2<R3<R1.

Table 2 Statistical table of chemical safety total score of GRAND

Reaction	Chemical	Flammability	Explosiveness	Toxicity	Reactivity

step	involved	Value	Score	$S_{FL(max)}$	Value	Score	$S_{EXP(max)}$	Value	Score	$S_{TOX(max)}$	Value	Score	$S_{r(max)}$
1	X1	-110	96.47	96.47	70.9	94.91	94.91			99.75	0	0.37	5.74
	X2	210	4.34								0	0.37	
	X3	221	3.52								0	0.37	
	X4	53	51.21		6.2	0.22		120	98.96		0	0.37	
	X5	215.5	3.91								0	0.37	
	X6	158.8	11.23								0	0.37	
	X7	200	5.26								0	0.37	
2	X8							2	99.75		1	5.74	
3	X8							2	99.75		1	5.74	

Note: X1, X2, X3, X4, X5, X6, X7, X8 represent hydrogen, 2-ethyl hydrazine, 2-ethylhydroquinone, heavy aromatic hydrocarbon, trioctyl phosphate, tetrahydro-2-ethyl Base, tetrahydro-2-ethylhydroquinone, hydrogen peroxide, respectively.

Table 3 Statistical table of process safety total score of GRAND

Reaction step	Temperature				Pressure			Heat of reaction			Process inventory		
	Value	Value -25℃	Score	$S_{T(max)}$	Value	Score	$S_{P(max)}$	Value	Score	$S_{PI(max)}$	Value	Score	$S_{HR(max)}$
1	50℃	25℃	0.33	0.33	3.45	1.33	1.33	-75.33	0.39	0.39	50	76.85	76.85
2	38℃	13℃	0.21		2.47	1.09		-29.22	0.29		92.5	1.98	
3	30℃	5℃	0.23		1.00	0.82		0	0		99.00	0.92	

Table 4 GRAND total score for H_2O_2 manufacturing process routes

Route	Chemical safety total score(CSTS)	Process safety total score(PSTS)	GRAND total score	GRAND ranking curve
R1	296.17	78.90	375.07	42
R2	187.09	24.77	211.86	6
R3	296.17	68.73	364.90	39

Note: R1, R2, and R3 represent the ruthenium method, the isopropanol method, and the direct synthesis of hydrogen and oxygen, respectively.

The score of R2 in GRAND ranking curve(Fig.2) has the lowest score (less than 10), indicating that it is safer than the other two routes. The ranking curves of the two routes R1 and R3 are 42 and 39, respectively, and the scores differ by only 3. From the perspective of the ranking curve, the difference in the risk scores between the two routes with similar scores is also very small.

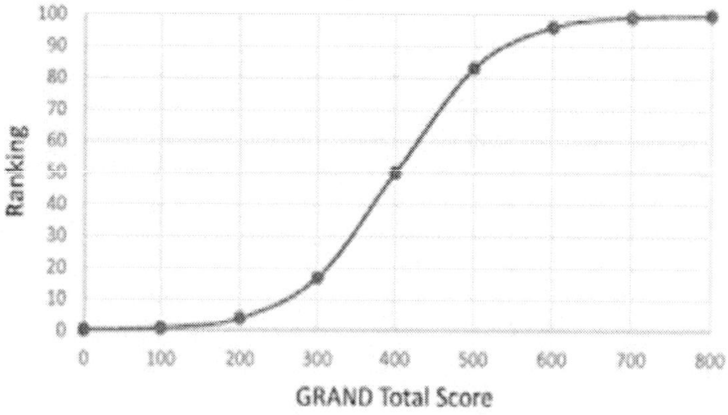

Fig. 2. GRAND ranking curve.

The GRAND ranking curve does not consider the root cause of the analytical method, nor does it have a specific mathematical model. The 2DGR is an extension of GRAND to further assess the hazard level of the process, as shown in Fig 3. Colour-coded regions provided by 2DGR will make it easier to distinguish between levels of hazardous process routes as compared to ranking as commonly used in the current inherent safety assessment methods.

The x-axis represents the total range of GRAND totals from 0 to 800, and the y-axis represents the most dangerous parameter. The range of MHP is 0-8. The GRAND total score is obtained by adding the scores of 8 parameters, and the score range of each parameter is 0-100. The number of MHPs represents the hazard level of the process route, and the more the number of processes, the more dangerous it is. The number of MHPs for each routing is determined by the highest score for each parameter in the different routing. 2DGR uses green, orange and red to indicate the minimum hazard, danger and maximum risk, respectively. Considering the ranking curve, the difference in hazard scores between R1 and R3 is small, and further analysis of which process route is more dangerous is needed. The results of the number of MHPs in each process route are shown in Table 5, and the evaluation results of individual parameters of GRAND are simplified.

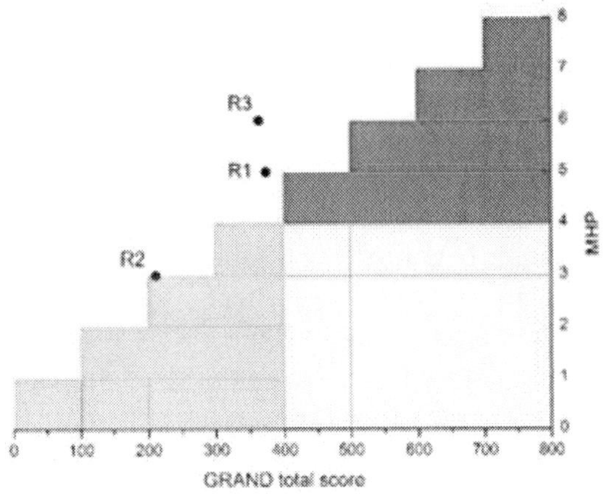

Fig. 3. 2 Dimensional graphical rating (2DGR) for inherent safety assessment hazard indicator.(red region- most hazard; yellow region - most hazard; green region - least hazard)

As seen from Table 5, the number of MHPs of R1, R2, and R3 are 5, 3, and 6, respectively. R1 contains 2 safest parameters (single parameter GRAND score is the lowest), respectively pressure and reaction heat, R3 also contains the two most safe parameters are temperature and reaction residual. Based on this evaluation, it is concluded that R3 is the most dangerous process route and R2 is the safest route. However, the difference in MHP between R1 and R3 is only 1, and the difference in hazard levels is also small, requiring further risk assessment.

Table 5 GRAND results for individual parameter assessment

Route	Number of the most hazardous parameter(MHP)	Type of hazardous parameter
R1	5	Flammability, Explosiveness, Toxicity, Reactivity, Process inventory
R2	3	Toxicity ,Reactivity, Process inventory
R3	6	Flammability, Explosiveness, Toxicity, Reactivity, Pressure, Heat of reaction

If the process hazard ranking is only considered from the single factor of GRAND ranking curve score and MHP quantity, the dangerous level of the route cannot be well judged. In the case of two process routes, GRAND total score, GRAND ranking curve, and MHP number are not much different, 2DGR can play its own advantages to identify the most dangerous process route in the graphics area, as shown in Figure 1. In Figure 1, R2 is in the green area, indicating a safe process route; the coordinate points of R1 and R3 fall in the upper part of the green area and in the white area near the red area, and fall in the specified 2DGR image. It is not possible to visually judge R1 and The dangerous level of R3. Due to the different numbers of MHPs for R1 and R3, the GRAND scores for the red danger zones where R1 and R3 enter the corresponding MHP are also different. Obviously, by adopting the principle of proximity, the GRAND score of R1 can be increased by 26.66 points to enter the red danger zone where the number of MHP is equal to 5. However, the GRAND score of R3 can be increased by 133.84 to enter the MHP. A red hazardous area with a number equal to 6. In summary, R1 is the most dangerous process route in the three hydrogen peroxide production processes.

3.2 The root-cause of hazards analysis

In order to determine why R1 is the most dangerous route, reverse analysis of R1 is performed. First, the reason why the R1 route becomes the highest GRAND total score is analyzed. The CSTS in R1 is 217.52 points higher than the PSTS. Therefore, it is determined that the quantitative score generated by some parameters in the CSTS is too high, resulting in an increase in the GRAND total score. Figure 2 is a GRAND analysis of the various parameter materials in the hydrogen peroxide production by the hydrazine process.

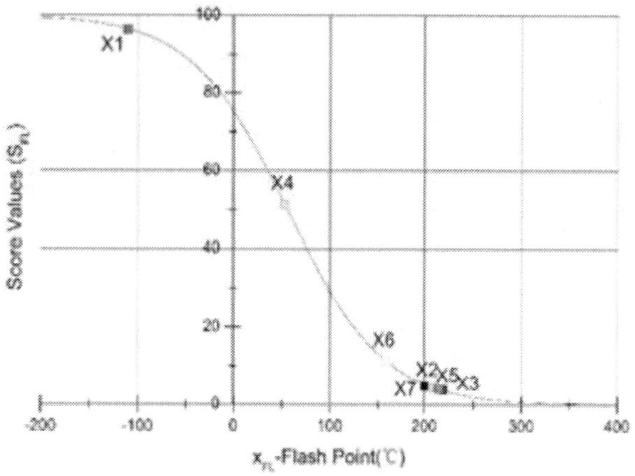

(1) GRAND Analysis of Flammability parameter-R1

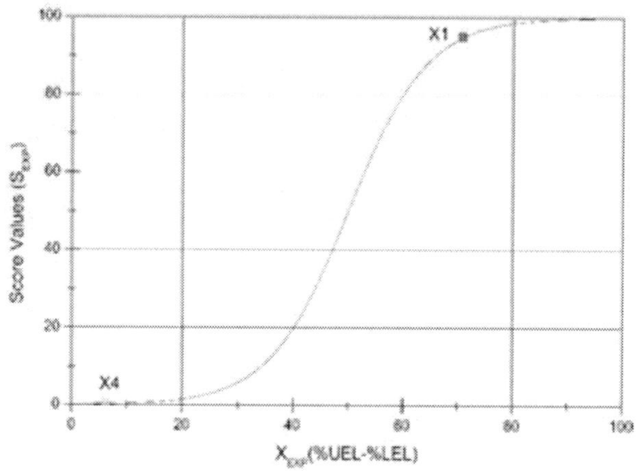

(2) GRAND Analysis of Explosiveness parameter-R1

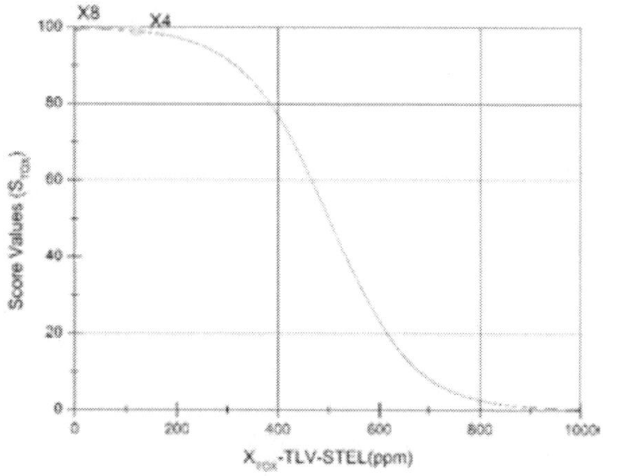

(3) GRAND Analysis of Toxicity parameter-R1

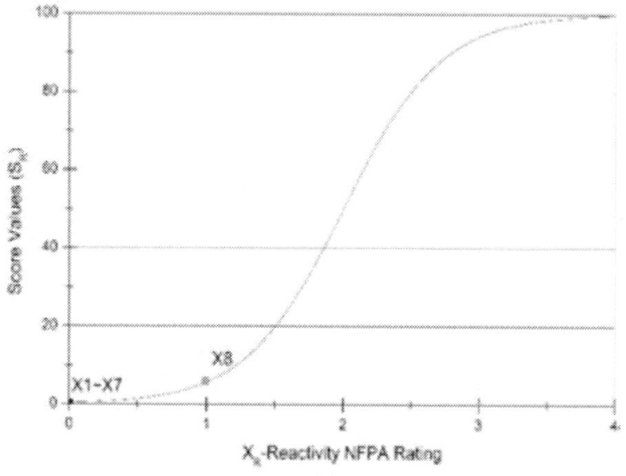

(4) GRAND Analysis of Reactivity parameter-R1

Fig. 2 GRAND analysis of various parameter -R1-based route

Note: X1, X2, X3, X4, X5, X6, X7, X8 represent hydrogen, 2-ethyl hydrazine, 2-ethylhydroquinone, heavy aromatic hydrocarbon, trioctyl phosphate, tetrahydro-2-ethyl Base, tetrahydro-2-ethylhydroquinone, hydrogen peroxide, respectively.

It can be seen from Fig. 2(a) that due to the lower flash point of hydrogen in S1, the GRAND score of the flammable parameter is higher, while the flash point value of other substances is distributed within 50-230 °C, which is relatively safe;

Fig. 2(b) shows that the broader the range of explosion limits, the more likely an explosion will occur. Hydrogen in a wide range of explosive limits is the direct cause of the high GRAND score, while the flammable liquid heavy aromatics have a narrow range of explosive limits due to low volatility, and the GRAND score for the explosion parameters is not significant.

Fig. 2(c) illustrates that chemicals with lower TLV-STEL values are more dangerous than chemicals with higher TLV-STEL values. When the short-time contact threshold exceeds 10-3, the GRAND fraction approaches zero. The chemical substances that cause high GRAND scores in S2 and S3 are hydrogen peroxide and heavy aromatics. Hydrogen peroxide is a strong oxidant. The vapor or mist inhaled by hydrogen peroxide is highly irritating to the respiratory tract. Irreversible damage or even blindness, heavy aromatics inhaled into the human body, causing pneumonia, and damage to the nervous system, liver, and skin. And it can be seen from Fig. 2(d) that the activity parameter of the mass fraction of 27.5% hydrogen peroxide does not produce a higher GRAND fraction, but the safety of hydrogen peroxide is particularly safe. Pure hydrogen peroxide is stable at any concentration. When it is in contact with heavy metals and their salts and alkaline substances under normal temperature and pressure, it is easy to cause an explosion reaction. According to the hazard rating index of NFPA in Table 1 [9], the hydrogen peroxide solution has a reactivity value of 4 when a heavy metal and its salts are encountered, and a GRAND score of 100, which reaches the maximum peak value. For example, when the Cu^{2+} concentration exceeds 0.04%, 30% of hydrogen peroxide begins to run out of control [10]. The increase in the reactivity of the hydrogen peroxide solution also increases the decomposition rate of the hydrogen peroxide solution, and the decomposition reaction rate increases as the temperature, pH and impurity content increase. For every 10 °C increase in temperature, the decomposition rate increases by about 1.3 times, and a large amount of oxygen and heat are released. The decomposition further accelerates the temperature rise and decomposition, forming a vicious circle, and at the same time CSTS and PSTS also occur correspondingly. Changes pose a great threat to production safety [11]. In both normal and abnormal situations, GRAND can quantify the risks in the hydrogen peroxide process and is of great significance.

4 Conclusion

The GRAND and 2DGR methods used not only allow quantitative risk assessment of the process, but also risk root cause analysis. Through the risk assessment of three processes for the production of hydrogen peroxide, it is determined that the production of hydrogen peroxide by helium is the most dangerous process route. Based on the root-source analysis of the hydrogen peroxide production route, the flammable and explosive parameters of hydrogen, the short-term toxicity contact threshold of hydrogen peroxide, heavy aromatics and the reactivity of hydrogen peroxide are the root causes of the process hazard.

5 References

[1] De-Jian, W. U. , Q. Xin-Ming , and H. Ping . "Research development of hydrogen peroxide thermal explosion." Journal of Safety Science & Technology (2011).

[2] Qi, Li , Z. Wenfang , and G. Yanxia . "Application of HAZOP in Hydrogen Peroxide Unit." Safety Health & Environment (2010).

[3] Mao Gang. "Application of Fire and Explosion Hazard Index Evaluation Method in Hydrogen Peroxide Production Unit." Chemical Management 3 (2014).

[4] Ebrahimi, F., T. Virkki-Hatakka, and I. Turunen. "Safety analysis of intensified processes." Chemical Engineering & Processing Process Intensification 52.52(2012):28-33.

[5] Ahmad, Syaza I., H. Hashim, and M. H. Hassim. "Numerical Descriptive Inherent Safety Technique (NuDIST) for inherent safety assessment in petrochemical industry." Process Safety & Environmental Protection92.5(2014):379-389.

[6] Ahmad, Syaza I. , H. Hashim , and M. H. Hassim . "A graphical method for assessing inherent safety during research and development phase of process design." Journal of Loss Prevention in the Process Industries 42(2016):59-69.

[7] Ebrahimi, F., T. Virkki-Hatakka, and I. Turunen. "Safety analysis of intensified processes." Chemical Engineering & Processing Process Intensification 52.52(2012):28-33.

[8] Rathnayaka, Samith, F. Khan, and P. Amyotte. "Risk-based process plant design considering inherent safety." Safety Science 70(2014):438-464.

[9] National Fire Protection Association., 2013. Retrieved 3.03.2013 from www.nfpa.org. Palaniappan, C., Srinivasan, R., Tan, R., 2002a. Expert system for the design of inherently safer processes. 1. Route selection stages. Ind. Eng. Chem. Res. 41, 6698-6710.

[10] Azatyan, V. V., et al. "On the possibility of a thermal explosion initiated by a heterogeneous reaction between H 2 and O 2 (comments on the paper “initiation of chain and thermal explosions by the reactor surface. criterion for the participation of branching chains in a therma." Combustion Explosion & Shock Waves 44.6(2008):732-736.

[11] Jiajian, Wang , and Z. Hong . "Safety Analysis and Prevention or Controlling Measures on Production Process of Hydrogen Peroxide by Anthraquinone Method." Safety Health & Environment (2012).

544288 Double Network Hydrogel from Acrylic Acid

Hydrogels attract great attention due to their soft and wet nature, similar to that of biomaterials. Recent innovations in several tough hydrogels show their potential as structural biomaterials, such as cartilage. In order for tough hydrogels to be used in biomedical applications, their required mechanical properties such stiffness, strength, toughness, damping, fatigue resistance and self-healing, along with biocompatibility are considered for every application. In this study, hydrogels were synthesized batch-wise with acrylic acid, methacrylamide, and poly-propylene glycol diacrylate. Sodium dodecyl sulfate was introduced to the polymer network to reduce the swelling, an issue presented in a water environment. Acrylic acid and methacrylamide are added followed by potassium persulfate in order to randomly co-polymerize the monomers at 75 °C. The resulting material displayed both hydrophilic and hydrophobic interactions, as well as a high tensile strength. The gel formations were physically characterized by Fourier Transform Infrared spectroscopy analysis and UV-Vis spectroscopy. Mechanical properties of the hydrogel were examined using shear rheology and tensile testing.

550493 Modeling Swelling Equilibrium of Sodium Polyacrylate in Water and Saline Water with Polyelectrolyte Non-Random Two Liquid Model

Superabsorbent polymers or hydrogels are widely used in daily hygiene product as well as multiple industrial applications. Proper modeling phase behavior of hydrogels helps improving the performance of the superabsorbent products. The swelling behavior of sodium polyacrylate, a typical type of superabsorbent, has been investigated experimentally by Ogawa and Lee, etc. Maurer used a modified Pitzer model to describe swelling equilibrium; Sadowski extended PC-SAFT model with two additional interaction terms to calculate VLE and water activity of polyelectrolyte hydrogels. Both models require extensive number of adjustable parameters and have limitation on description of speciation in the hydrogel phase.

Based on our recently developed electrolyte non-random two liquid model, swelling equilibrium between sodium polyacrylate hydrogel and saline water of different concentrations has been analyzed. Phase equilibrium is modeled according to equal chemical potential of mobile ion components with a correction factor to account for the elastic network of the polymer. The speciation and activity coefficients of mobile ions in hydrogels can be found with this thermodynamic framework for the first time. Model prediction for degree of absorption is in good agreement with available data.

1. Ogawa, I., Yamano, H., & Miyagawa, K. (1993). Rate of swelling of sodium polyacrylate. *Journal of applied polymer science*, *47*(2), 217-222.
2. Lee, W. F., & Huang, Y. L. (2001). Superabsorbent polymeric materials X: Effect of degree of neutralization on swelling behavior of crosslinked poly (sodium acrylate) in aqueous salt solutions. *Journal of Polymer Research*, *8*(1), 9-15.
3. Thiel, J., & Maurer, G. (1999). Swelling equilibrium of poly (acrylamide) gels in aqueous salt and polymer solutions. *Fluid phase equilibria*, *165*(2), 225-260.
4. Arndt, M. C., & Sadowski, G. (2014). Thermodynamic model for polyelectrolyte hydrogels. *The Journal of Physical Chemistry B*, *118*(35), 10534-10542.

550563 Thermodynamic Modeling of Aqueous Multivalent Polyelectrolyte Systems with Polyelectrolyte NRTL Model

Polyelectrolytes are a type of commonly seen polymers both in human body and in industry. Some examples are DNA and RNA in living cells, selective permeable membranes, and ion-exchange resins. Specific features of polyelectrolytes such as hydrophilic and hydrophobic segments along the polymer backbone endow these molecules with unique and intriguing behaviors.

A recent study reports a novel polyelectrolyte NRTL model for thermodynamic modeling of aqueous monovalent polyelectrolytes [1]. The model expresses the excess Gibbs energy of the polyelectrolyte solutions as the sum of two terms: long-range interaction term and short-range interaction term. The long-range interaction term accounts for the charged polyion segment-ion and ion-ion interactions with the Pitzer-Debye-Hückel limiting law [2] and the polyion backbone-ion interactions with the Manning limiting law [3]. The short-range interactions, including molecule-molecule, molecule-ion/polyion and ion/polyion-ion/polyion interactions, are treated with the electrolyte non-random two-liquid (eNRTL) equation [4]. With use of binary interaction parameters, the model has been shown to successfully correlate experimental data of various colligate properties, such as osmotic coefficients of the polyelectrolyte solutions, activity coefficients of counterions, and mean ionic activity coefficients of added salts, for aqueous monovalent polyelectrolyte solution [1]. In this study, we review literature thermodynamic data for aqueous divalent and multivalent polyelectrolytes and highlight the differences from monovalent ones. We further examine the polyelectrolyte NRTL model behavior for aqueous multivalent polyelectrolyte systems and show successful correlations of experimental activity coefficient data for multivalent counterions and added salts, and osmotic coefficient data for multivalent polyelectrolyte solutions. Future studies should involve extension of the polyelectrolyte NRTL model for aqueous mixed polyelectrolyte solutions.

Reference:

[1] Yu, Y., Hossain, N., Chen, C.-C. Modeling of Polyelectrolytes System with Electrolyte Nonrandom Two Liquid Model. AIChE Journal. In preparation.

[2] Pitzer, K. S. Thermodynamics of Electrolyte. I. Theoretical Basis and General Equations. The Journal of Physical Chemistry. 77 (1973) 268-277.

[3] Manning G. S. Limiting Laws and Counterion Condensation in Polyelectrolyte Solutions I. Colligative Properties. The Journal of Chemical Physics. 51 (1969) 924-933.

[4] Song, Y., Chen, C.-C. Symmetric Electrolyte Nonrandom Two-Liquid Activity Coefficient Model. Industrial & Engineering Chemistry Research. 48 (2009) 7788-7797.

550668 Thermodynamic Modelling of Ca^{2+}-Na^{+}- Cl^{-}-CO_2 Quaternary System with Electrolyte NRTL Model

Calcium carbonate ($CaCO_3$) is one of the most important solutes that can lead to scale formation in produced water processing equipment and in desalination water treatment. A scientific model that provides quantitative thermodynamic predictions for different conditions is useful and when used with models of the thermodynamics of other important ions can form a complete model of scaling and precipitation in such equipment. In this study symmetric electrolyte nonrandom two liquid (eNRTL) activity coefficient model is used to model the aqueous Ca^{2+}-Na^{+}- Cl^{-}-CO_2 quaternary system. The model requires only molecule-electrolyte and electrolyte-electrolyte binary interaction parameters. The temperature dependence of these interaction parameters is described via Gibbs-Helmholtz equation including up to three temperature coefficients. Experimental data including the osmotic coefficient, mean ionic activity coefficient, vapor pressure, heat capacity, excess enthalpy, and salt solubility at different temperatures and electrolyte concentrations up to saturation are used to identify the model parameters. The model provides accurate representation for all thermodynamic properties and phase equilibria for the aqueous quaternary system. The model of carbonate thermodynamics is coupled with the thermodynamics of other important electrolytes to evaluate scaling and precipitation in typical saline waters including brackish groundwater, seawater and examples of produced waters.

551178 Improved Oil Recovery with Viscoelastic Surfactant: Screening and Effectiveness of Formulations

Middle East carbonate reservoirs are usually characterized as high temperature and high saline (HTHS) reservoirs. In chemical enhanced oil recovery (cEOR), interfacial tension (IFT) and capillary number are the key parameters that alter the reservoir environment. Viscoelastic surfactant (VES) is introduced due to its dual capability of lowering the IFT and improving the sweep efficiency of the reservoir. In this work, the main objective is to screen out a novel VES for cEOR applications in HTHS carbonate reservoirs.

Novel VES solutions of 0.1 to 1 wt% concentrations were prepared with synthetic sea water and stirred until they attain equilibrium. Densities and IFT's of VES solutions and oil at different temperatures (30°C, 50°C and 80°C) were determined using density meter and spinning drop tensiometer (Kruss). Equilibrated solutions were used with heavy crude oil of 17°API to determine the IFT. Both short term and long term thermal stability of VES was evaluated using an approach based on FTIR, NMR and TGA analyses. Aging of VES was performed for 30 days at two high temperatures of 90°C and 120°C.

IFT results showed considerable reduction in IFT with novel VES and heavy oil upto 10^{-1} mN/m range. IFT decreased initially and then increased with increase in VES concentrations. Non-linear trend of IFT was obtained. IFT reduction usually depends on the concentration of solutions used. Results also indicate that ultralow IFT was obtained at 50°C. Critical micelle measurements were performed and 3wt% was found as CMC. VES was evaluated for short term thermal stability using TGA analysis and found to be stable upto 250°C. FTIR, ^{1}H NMR and ^{13}C NMR before and after aging the VES samples were compared and structural changes were observed in the aliphatic region. The peaks were shifted away due to the presence of C-O bond in the sample. IFT for aged samples was determined and almost two folds of increase in IFT was observed. Depending upon the results of this investigation at high temperatures from FTIR, ^{1}H NMR and ^{13}C NMR and IFT, this novel VES exhibit great potential to be used in cEOR applications when we have HTHS reservoirs.

Based on the ultralow IFT's and thermal stability obtained with the viscoelastic surfactant, they are excellent additives that can be used in chemical EOR for HTHS conditions. They are also very efficient when we have heavy oil reservoirs and where IFT reduction is required in order to mobilize the oil and improve the sweep efficiency of the reservoir.

551181 Laboratory Investigation of Viscoelastic Surfactant for Chemical EOR in High Salinity and High Temperature Carbonate Reservoirs

Viscoelastic surfactants are gaining much attention in the recent years due to their structure. Acquiring dual capabilities of a polymer as well as a surfactant in one composition provides considerable value to chemical EOR. Middle East reservoirs are characterized as high salinity and high temperature that make carbonate reservoirs in this region even more challenging. In this work, a systematic experimental study is carried out to test the applicability of novel viscoelastic surfactant (VES-AGA 97) in enhancing the oil recovery.

Rigorous investigation of VES was performed to determine its potential in improving oil recovery. Screening of VES was carried out using five major evaluation techniques. Thermal stability of VES was determined using TGA and FTIR methods. Interfacial tension (IFT) measurements were recorded using spinning drop tensiometer. ^{1}H NMR and ^{13}C NMR were measured with long-term thermal stability tests performed for 30 days in anaerobic conditions at two different high temperatures. Static adsorption was measured by shaking the crushed powder samples with VES for 24 hours and measuring the changes in concentration of VES. Dynamic adsorption is estimated by performing core flooding experiments at 90°C.

Experimental results with TGA demonstrated that VES was thermally stable and showed high resistance to temperatures up to 250°C. FTIR results indicated good thermal stability with strong bonding between C-H and C=O. Considerable reduction in IFT with VES up to 10^{-1} mN/m range was observed. Effect of concentration, time and temperature on IFT was evaluated. It was also noted that ultralow IFT was obtained at 50°C and 80°C. ^{1}H NMR showed stability in structure while ^{13}C NMR results showed minor structural changes in the aliphatic region. Static adsorption results gave less adsorption when we had high temperatures compared to low temperatures. Langmuir and BET adsorption isotherm models were utilized to determine maximum adsorption rate. Comparison between both models and experimental results exhibited perfect match. Dynamic adsorption was evaluated using concentration analysis of effluents collected at the downstream from core flooding. Difference between the concentrations of VES at inlet and outlet was measured using total organic carbon analysis. Additional oil recovery of 8% was determined when recovery factor is plotted against pore volumes.

The novelty is the integrated approach of screening and application of viscoelastic surfactant for chemical EOR applications in improving the oil recovery. Thermal stability, ultralow IFT's, adsorption and additional oil recovery obtained with viscoelastic surfactant showed that they are excellent additives that can be used in chemical EOR for high salinity and high temperature conditions. They are also very useful when we have heavy crude oil and where IFT reduction is required to mobilize the oil and improve the sweep efficiency of the reservoir.

550460 Ionic Transport in Reverse Osmosis Membrane

Nano-filtration membrane (NF)/ reverse osmosis membrane (RO) can be described by extended Nernst-Planck equation, where separation is a result of diffusion, convection and electro-migration of ionic species. Poisson-Boltzmann equation is used to describe potential profile. Ionic species are electro-neutrality everywhere in bulk of feed channel, and potential of solution is zero in this region. In electrical double layer region where is close to the charged membrane surface, negative static charged trends to repel co-ions and attracts counter-ions, and distinct ionic concentration distribution is formed. Co-ions and counter-ions would be balanced inside membrane, where potential change is assumed to be linear. We assume electro-migration effect in permeate site is minor. It is noticed from the numerical solution that the rejection of chloride was lower than the rejection of sodium throughout membrane. The rejection of both ions increased as the volumetric permeating flux based on membrane area increased. An extension to multiple species shows that fixed charged membrane has a priority to reject divalent, sulfate, ions more than monovalent, chloride .

549541 A Brief Review of Recent Development of Process Safety Management for Chemical Industry in East Asia

Since the chemical production of China overtook that of US in 2008/2009, the total chemical production and exports in East Asia has increased rapidly since 2000s. Now, China, Korea, Japan and Taiwan combined to export more than one quarter of tops 15 countries (with 78%) of the world. Those prosper development had also brought about some severe unfortunate incidents due to miss management. There were many new laws of process safety management enacted since 2000s or even recently—along with the newly establishment of national /local emergency response measures after 2010. Some results were observed as the incidents rates decreased. However, more fundamental changes will be needed to cover the old infrastructures, such as pipelines, old petrochemical plants…etc. In addition, the laws and actions to avoid industrial grade chemicals to be used as food additives or entering the food process will be the food safety issues for the years to come.

551220 Advanced Two-Tip Optical Fiber Probe Technique to Local Gas and Liquid Saturation and Velocities Measurement within Various Commercial Catalyst Shapes in Pilot Plant Trickle Bed Reactors (TBRs)

Advanced two-tip optical fiber probe (OFP) was manufactured and developed in Multiphase Flows Reactor and Engineering and Applications Laboratory (mfReal) in Missouri University of Science and Technology. The OFP has been implemented to study vital hydrodynamic parameters in multiphase reactors and flow systems. OFP is promised techniques to assess different dynamic flow parameters (channeling, bypassing, phase distribution, catalyst utilization, and local phase velocity). Plexiglas transparent column with 14 cm inside diameter and 200 cm height packed within different industrial catalyst shapes (Trilobe, Quadralobe, Spherical, and Cylindrical) and wide range of gas and liquid velocities. The acquired data by OFP were validated with the data obtained by a syringe pump and x-ray digital industrial radiography (DIR). Local gas and liquid saturation represent a crucial parameter in TBRs which play a central role to predicate most the dynamic flow parameters in modeling and correlations and supply window to track the phase distribution along the height bed with diameter profile as well. Meanwhile, provide a high contribution by measuring local gas and liquid velocities which represent in high demand information for industry to support and improve heat and mass transfer coefficient study, modeling and CFD simulation as well. In the current study, the results have been showed a high capability and high reliability for OFP to measure dynamic flow parameters (local gas and liquid saturation and velocities) within different catalyst shapes. Also, a significant impact of catalyst shapes on the hydrodynamic parameters in pilot plant TBRs.

551261 Morphology Evolution in Titanium Dioxide Nanotubes Arrays Prepared By Electrochemicalanodization

The production of hydrogen by electrochemical dissociation of water has been known for minimal environmental impact involved. This process consumes more electrical energy than electricity generated by combustion of produced hydrogen, but it is still interesting in order to find sustainable future sources of clean energy. One way to address this problem is to investigate options for reducing inefficiencies on the electrochemical cell, these are voltage consuming, it is also possible to take advantage of cheaper energy sources. Photocatalysis has established as viable option in the development of processes for the treatment of pollutants and clean energy production. This option is based on the ability of semiconductors to generate an electron flow by means of the interaction with solar radiation.

Owing to its electronic structure, TiO2 is the most frequently used semiconductors in photocatalysis, although it has a high recombination of photogenerated charges and low solar energy absorption. Due to TiO2 can produce electricity from solar energy, it can be used to satisfice the electrical energy requirements on the operation of electrochemical cells. During this process pairs hole – electron are photogenerated, the efficiency can be improved if charge recombination is avoided, also if the energy absorption into the visible spectrum is increased and its specific surface area is increased. An alternative to reduce these limitations is the use of nanostructured morphologies which can be produced during the synthesis of TiO2 nanotubes (TNTs). Therefore, if possible to produce oriented nanostructures it will be possible to generate a greater contact area with electrolyte and better charge transfer. At present, however, the development of these innovative structures still presents an important challenge for the development of competitive photoelectrochemical devices.

This research focuses on established correlations between synthesis variables and nanostructure morphology which has a direct effect on the photocatalytic performance. TNTs with controlled morphology were synthesized by potentiostatic anodization of titanium foil. The anodization was carried out at room temperature in an electrolyte composed of ammonium fluoride, deionized water and ethylene glycol. Consequent thermal annealing of as-prepared TNTs was conducted in the air at 450 °C. Morphology and crystalline phase of the TNTs were analyzed by SEM, EDS and XRD. As results, the synthesis conditions were established to produce nanostructures with specific morphological characteristics. Anatase was the predominant phase of TNTs after thermal treatment. Nanotubes longer than 7 μm and pore diameter around 50 nm with different surface-volume ratio are important in photoelectrochemical applications based on TiO2 due to their dimmensional characteristics, surface-volume ratio, reduced radial dimensions and high oxide/electrolyte interface. Finally, this knowledge can be used to improve the photocatalytic activity of TNTs by making additional surface modifications with dopants that improve their efficiency.

551360 Synthesis, MR Relaxivities, and in Vitro Cytotoxicity of Ligands Coated Gadolinium Oxide Nanomaterials

Magnetic resonance imaging (MRI) is widely used in modern clinical medicine as a diagnostic tool, and provides noninvasive and three-dimensional visualization of biological phenomena in living organisms with high spatial and temporal resolution. Therefore, considerable attention has been paid to magnetic nanoparticles as MRI contrast agents.

We report a facile method to synthesize high quality and bio-functionalized Gd_2O_3 nanoparticles (BFNPs) for use as contrast agents in MRI. The bonding status of BFNPs were confirmed by FT-IR and TGA analysis. The surface coating amount was estimated to be from 40% to 60% in weight percent from a TGA analysis. High voltage electron microscope (HVEM) shows that the BFNPs were spherical in shape with an average diameter 3 mM. In addition, the bio-compatibility of the nanoparticles were measured by cytotoxicity tests by using human prostate cancer (DU145) and normal mouse hepatocyte (NCTC1469) cell lines which indicated that BFNPs are not toxic up to 250 mM. BFNPs are paramagnetic but have an appreciable magnetic moment at room temperature. This is because Gd(III) has seven unpaired 4f-electrons (S = 7/2). Therefore, appreciable r_1 and r_2 values are expected from sample solutions, which were in fact observed in this study. The r_1 and r_2 values of BFNPs were estimated to be 13.77 to 64.14 s^{-1} mM^{-1} respectively. The high relaxivities provide an opportunity to conduct perfusion MRI experiments with significantly lower concentrations than those needed for current commercial agents. A pronounced positive and negative contrast enhancement was clearly observed in 3 tesla T_1 MR images of a rat with a liver tumor after injection of an aqueous sample solution into a rat tail vein

553581 Analysis of Polar Compounds in Essential Bio-Oils By Fourier Transform Ion Cyclotron Resonance Mass Spectrometry

Analysis of Polar Compounds in Essential Bio-oils by Fourier Transform Ion Cyclotron Resonance Mass Spectrometry

Six essential oils obtained of Colombian species were analyzed by Fourier transform ion cyclotron resonance mass spectrometry coupled to negative electrospray ionization (ESI(-)/FT-ICR-MS), technique offered a selective ionization of non-basic nitrogen compounds and acid compounds[1]. Only deprotonated molecular ions or molecular ion adducts (Na^+ or Cl^-) are observed in the mass spectrum the results (figure 1) showed that the main classes were detected in the most bio-oils (classes with a percentage higher than 1, in relative abundance), but each class presents a different distribution and compounds. Have been identified between 1100-3600 compounds in the samples studied that present between 1-36 atoms of carbon. The most dominant classes were the oxygen compounds O4, O2, and O3. Found different and unique compounds in each bio-oil.

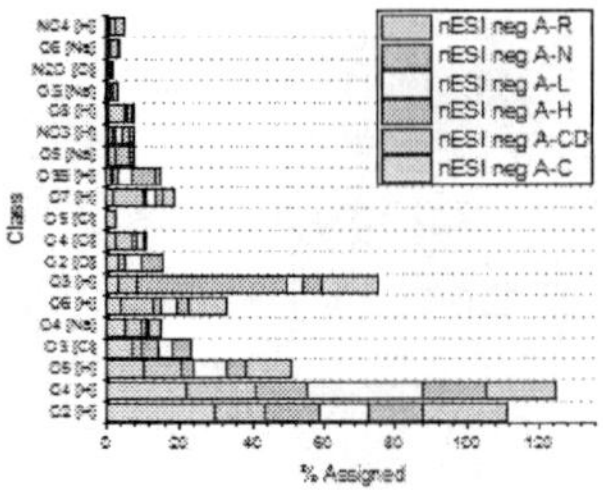

Figure 1. Principal classes detected by ESI(-) FT-ICR-MS of six essential bio-oils

[1] Arenas-Diaz, J. P. et al. Chemical characterization of polar species in Colombian vacuum residue and its supercritical fluid extraction subfractions using electrospray ionization FT-ICR mass spectrometry. Chem. Eng. Trans. 57, (2017).

549625 Influence of Batchelor Flow on Agglomeration of NMC Hydroxide

Influence of Batchelor Flow on Agglomeration of NMC hydroxide

Xiao-Tong Sun [1], Woo-Sik Kim[*, 1]

[1] *Department of Chemical Engineering, college of engineering, Kyung Hee University,*

Youngin, 17104, Korea

Abstract

In manufacturing the cathode material of lithium battery, morphology, size and size distribution of NMC oxide are directly related to the packing efficiency, which influence its energy density in battery performance. For a high packing efficiency of battery, spherical and size-uniform NMC hydroxides are thus required, which is determined by agglomeration process in reaction crystallization. [1-2] So, understanding and investigating the agglomeration process is necessary to obtain the desired agglomerates of morphology, size and size distribution. [1-4] Therefore, the present study focused on the influence of periodic fluid fields of Batchelor flow on agglomeration of NMC hydroxide.

The Batchelor flow induced between two concentric disks of crystallizer (called Rotation disk (RD) crystallizer) was applied for the reaction crystallization of NMC hydroxides, as shown in Fig.1 (a). Due to effective collision of particles and shaping process in periodic Batchelor flow, the spherical and size-uniform agglomerates of NMC hydroxides was produced in RD crystallizer, which was operated in continuous mode. Here, the mean residence time of RD crystallizer was 0.5hr, which was much shorter than in any previous studies running in mixing tank crystallizers at batch or continuous operations. A high fluid shear of Batchelor flow reduced the mean particle size and changed the irregularly shaped agglomerate particles into spherical ones, also, the agglomerate size distribution was narrower with increasing rotation speed of inner disk. (Fig.1 (b)) As a result, the agglomerates of NMC hydroxide with the tap density of about 2.1 g/cm^3 were obtained at a high rotation speed of 1500 rpm in RD crystallizer.

So, it could be inferred that that a RD crystallizer was highly applicable for the practical production of the agglomerate of NMC hydroxide and more effective than the conventional mixing tank crystallizer.

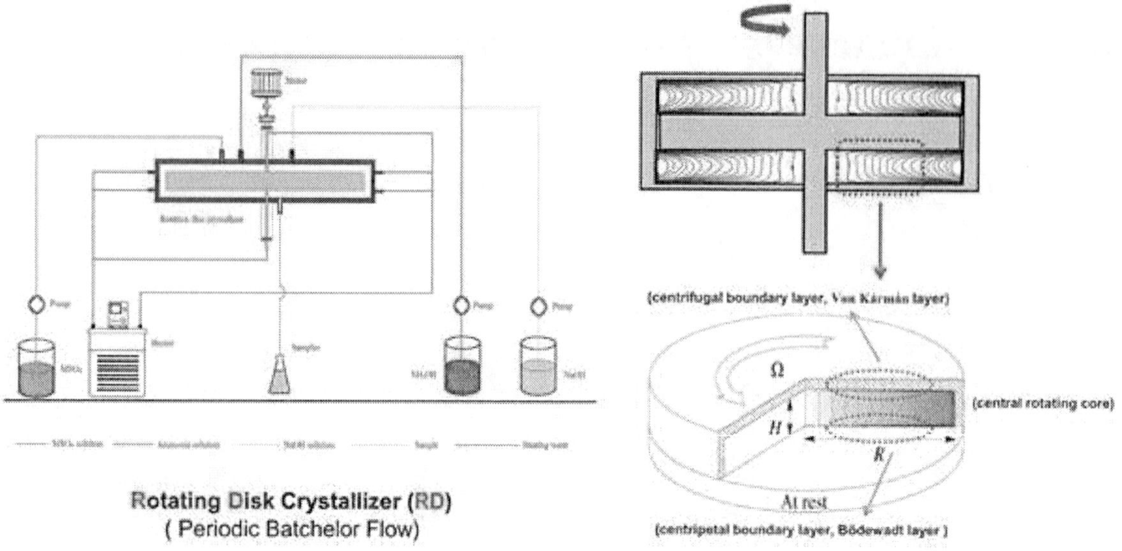

(a)

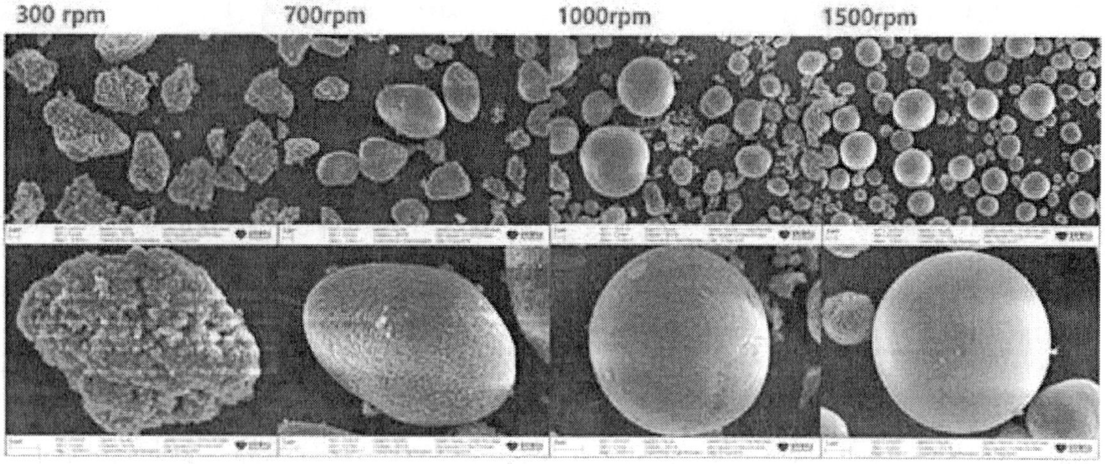

(b)

Figure1. (a) Reaction system & Batchelor flow in Rotation disk crystallizer, (b) agglomerates of NMC hydroxide at various rotation speed of crystallizer.

References

[1] Q. P. Mayra, W. S. Kim, *Cryst. Growth Des.*, 2015, 15 (4), pp 1726¨C1734

[2] J. E. Kim, W. S. Kim, *Cryst. Growth Des.*, 2017, 17 (7), pp 3677¨C3686

[3] J. M. Kim, S. M. Chang, J. H. Chang, W. S. Kim, *Collo. Surf. A: Physicochem. Engineering Aspects.*, 384 (2011) 31-39

[4] D. K. Thai, Q. P. Mayra, W. S. Kim, *Powder Technology.*, 274 (2015) 5-13

550343 Possible FIRE Scenarios in Electrical Substations Due to Formation of Flammable Mixtures

This study investigates the possibility of the occurrence of fire and explosion accidents caused by the vaporization of hydrocarbon components derived from mineral oil (dielectric Fluid) which is used as transformer insulating fluid in electrical substations. Samples of mineral oil have been collected from one of the electrical substations in Riyadh, Saudi Arabia. The samples were analyzed using Gas Chromatography Mass Spectrometry (GC-MS) to identify its composition. The results obtained from the GC-MS indicated that, 33 components varying from C_8 to C_{35} were present in the sample. Results revealed that the majority of these components are categorized as Alkane, Alkene, and Alkyne groups with some species from the Ketone, Alcohol, Aromatics, and Anhydride groups. About 25 % of the species were from Alkane group.

Mole fractions of the components presented in the liquid phase were obtained from the GC-MS results. The stoichiometric concentration method was used to calculate the Lower Flammability Limits (LFL) and the Upper Flammability Limits (UFL) for each component. The LFL and UFL for the mixture (LFL_{mix} and UFL_{mix}) were calculated in accordance to Le Chatelier equation. The LFL_{mix} and UFL_{mix} were found to be 0.88 vol. % and 5.75 vol. %. Then flammability diagram was used to determine if a flammable mixture existed. Results indicate that the vapour mixture for the sample is not flammable at 25°C and 1 atm. However, the vapour mixture was found to be flammable at 500°C.

553690 Continuous Power Production from Salinity Gradients Using Concentration Flow Cell Stacks

Salinity Gradient (SG) energy technologies are powered by the Gibbs free energy that is released from the mixing of two solutions with different salinities. This renewable energy source exists worldwide (e.g., in places like river deltas and industrial sites), and investigation of the methods that utilize this energy source have been focused on improving both the economic and electrical efficiency in order to make subsequent commercialization possible. Recently, concentration flow cells (CFC) have been introduced as a new technique with the highest reported power density for SG technologies so far. This improvement is the result of the utilization of both the Donnan potential from Cl- activity across the membrane as well as the electrode potential from Na+ intercalation/deintercalation processes. In this study, multiple CFCs were stacked together with a continuous flow design in order to further increase the power output. Experimental investigation was carried out for a consecutive number of stacks (1-5) using both parallel and series electrical wire connections. The parallel 5-stack exhibited superiority with a 3.1-fold increase in the total power output (8.6 mW) and a 4.6-fold increase in the absolute current (76.1 mA) when compared to the single cell (2.8 mW and 16.6 mA). The series stack exhibited a performance loss which was determined to be the result of an ionic-cross conduction, a phenomenon that has previously been reported in proton exchange membrane fuel cell stacks and microbial fuel cell stacks. Nevertheless, an effective increase in the overall power output with an increase in stack size was observed for both stack configurations. Additionally, the realization of the ionic-cross conduction from the series electrical connection is a key implication for designing a successful CFC stack for commercialization in the future.

546409 Performance Evaluation of Dust Suppressors of Ore Storage Piles By Spectroscopy and Wind Tunnel

The emission of fine particulate matter as dust from iron ore stockpiles is one of the concerns within logistics in the mining industry. In addition to possible environmental and health damages, iron ore being the raw material for the steelmaking process, economic losses point to the problem. The use of dust suppressors sprinkled on the surface of iron ore stockpiles is an effective method to reduce the emission of particulate matter, however due to the instability of the system: ore surface and suppressor agent there is no consensus regarding the expected properties of an efficient suppressor. Here we have developed a methodology to evaluate the performance of dust suppressors in iron ore piles using wind tunnel. Furthermore here is also proposed a test for prediction of suppressor performance using multivariate analysis with infrared spectroscopy data of suppressors with 86% accuracy. It was verified that the suppressors performance does not correlate linearly with the surface tension, wettability property. The suppressors tested did not significantly contaminate the iron ore confirmed by the thermal analysis of the samples. The high performance suppressors showed the hydrophobicity index of the bands 2926/3300 cm^{-1} lower than 0.333. In addition, this methodology can be adapted for the evaluation of suppressors of particulate material of other types of materials that demand bulk storage, in order to assist in the evaluation of the best products.

549111 Carbon-Based Au Catalyst for Acetylene Hydrochlorination Process: From Laboratory Data to Industrial Verification

Acetylene hydrochlorination is the main reaction for the dominant coal based process for VCM (vinyl chloride monomer) production from coal resource in mainland China. The traditional activated carbon based mercury catalyst for this reaction causes severe human damage and environmental pollution. Gold is widely considered as the most active metal for acetylene hydrochlorination reaction and copper is the most common metal with cost-effective advantage. The combination of two metals shows the best potential as the active component for environmentally friendly non-mercury catalysts. In our study, the performance of Au/Cu catalyst was enhanced dramatically by the introduction of organic ion ligands. Several ligands including trichloroisocyanuric acid (TCCA), cyanuric acid (CA) and melamine (M) were studied and TCCA was proved to be the best ligand for preventing Au^{3+} from reduction by acetylene, leading to longer catalyst lifetime. The findings were successfully verified by over 18000 hour's industrial plant-trial evaluation, and the VCM productivity reached about 4.2 times comparing to industrial mercury catalyst. Abundant characterization methods including TEM, XRD, TPR, XPS, ICP-AES and nitrogen adsorption-desorption were carried out to reveal the factors which affect the activity and stability of catalysts. Meanwhile, DFT calculation results also illustrated the stability of catalyst is strongly related to the adsorption energy of reactants on the surface of catalysts. As a conclusion, TCCA can strongly interact with Au^{3+} active sites, reducing acetylene adsorption energy and inhibiting the reduction of high valence Au by acetylene, thus enhance the stability of the catalysts. Furthermore, the electrophilic atoms and conjugated molecular structure of TCCA are important for the stabilization effect. The experiment and DFT results provide reasonable methodology for the design of high performance non-mercury catalysts in acetylene hydrochlorination process, which demonstrate enormous potential for the industrialization in PVC (poly vinyl chloride) industry.

550734 Novel Method Development for Pfas Contaminants Treatment in Landfills Leachates

This paper will present a novel approach that combine with sorbent and chemical reduction for the selective removal and destruction of per- and polyfluoroalkyl substances (PFAS) from landfill leachate. Although the co-contaminants, low concentration, and complexity for PFAS in industrial wastewater are big hurdles for current technologies. this method will overcome the challenges and provide a unique approach for PFAS removal.

The advantages of this technology include:
1) Capable of handling co-contaminants;
2) High loading flow rate;
3) Energy-efficient re-generable with PFAS recovery for destruction.

This method has the potential to reduce the combined PFAS concentration to below 70 ppt for wastewater treatment market, which is growing 19% annually to $11 billion until 2025. The technology is cost-effective and compatible with other treatment processes.

551186 Recyclable Plastic Resins Production from Waste Vegetable Oil and Biomass

Recently the Plastics Division of the American Chemistry Council (ACC) announced three goals designed to accelerate the industry's transition to a more circular economy [1]: 1) 100% of plastics packaging will be re-used, recycled or recovered by 2040; 2) 100% of plastics packaging will be recyclable or recoverable by 2030; and 3) 100% of the U.S. manufacturing sites operated by ACC's Plastics Division members will participate in Operation Clean Sweep-Blue by 2020. This paper will present a novel method for producing renewable epoxy polymer resins from waste vegetable oil and biomass-derived 2,5-furandicarboxylic acid. The epoxy polymer resins can be used as recyclable plastics, composites recycling and reusable 3-D printing filaments.
[1] Rick Lingle, Plastic resin producers and Circular Economy Goals: An interview with the ACC, Packaging Sustainability, Recycling, Business, Materials, May 29, 2018.

551320 N-Doped Cu/Cu_xo/C Foam Catalyst Having Multi-Crystaline Derived from Alkaline N-Coordinated HKUST-1/CMC for Enhanced 4-Nitrophenol Reduction

N-doped Cu/Cu_xO/C foam catalysts with multi-crystalline are synthesized from N-coordinated HKUST-1/cellulose for enhanced 4-nitrophenol reduction. Effect of N content and basicity on Cu morphology, crystal lattice and size, component dispersion and oxidation states in catalysts are systematically investigated. The facility of nano-catalyst synthesis from crystalline powders to robust morphology based foamed catalysts strategy is confirmed crediting to its simplicity and industrial applications. It is revealed that (1) graphitic nano-structure has been detected from carbonized N-doped HKUST-1 due to the catalysis of Cu sites towards C/N elements in HKUST-1; (2) alkaline N-doping inhibits the crystal growth of Cu species, resulting in crystal size of 3-5 nm in porous catalyst samples; (3) DFT simulation verifies that N doping will lower HOMO and LUMO energies of Cu-BTC in various degree It significantly restrains a complete reduction of Cu ions in HKUST-1, and facilitates the formation of a multi-crystalline Cu/Cu_xO/C based from HKUST-1 calcination; (4) amount of Cu_2O in N-doped Cu/Cu_xO/C gradually increases with the enhanced alkaline strength and content of the N dopant in HKUST-1. Catalysis performance exhibits that our designed N-doped Cu/Cu_xO/C based from HKUST-1s show much higher fast catalysis kinetics and better recycle performance on par with the start-of-the-art catalysts. Among which, melamine-doped Cu/Cu_xO/C foam catalyst exhibits the most outstanding activity in the reduction of 4-nitrophenol with a kinetic rate of 1.75 min^{-1}, 21 times faster comparing with bare Cu/C catalyst based on pristine HKUST-1, demonstrating the critical contribution of the foamed morphology and multi-crystalline of Cu/Cu_xO/C to the outstanding activity and recycle performance.

555799 Climate Solutions Presentation

This electronic poster will discuss the update of AIChE's newest Climate Solutions group.

543923 Dissolved Oil Removal from Ethylene Plants Quench Water

A method for removing dissolved hydrocarbons from water may comprise: cracking a mixed hydrocarbon stream in a cracking furnace to produce a cracked gas effluent; quenching the cracked gas effluent in a quench water tower with quench water to produce a quenched gas stream and a spent quench water stream comprising water, tars, heavy aromatic hydrocarbons, gasoline, dissolved oil, and dispersed oil; feeding the spent quench water stream to a liquid-liquid extraction unit wherein the liquid-liquid extraction unit removes at least a portion of the dissolved oil and produce an extracted effluent stream.

549676 NOVA Chemical's Ethylene Plant 18 Year Continuous Operations with Chevron Phillips Chemical Company's E-SERIES® Catalyst

NOVA CHEMICAL'S ETHYLENE PLANT 18 YEAR CONTINUOUS OPERATIONS WITH CHEVRON PHILLIPS CHEMICAL COMPANY'S E-SERIES® CATALYST

Maite Sadaba Omana, E3 Contact Engineer, NOVA Chemicals; Steve Frezell, E3 Operations Coordinator, NOVA Chemicals; Kelly Pritchett, E-Series® Catalyst Technical Sales and Service Representative, Chevron Phillips Chemical Company LP

This paper will discuss the operating discipline and trends associated with the Chevron Phillips Chemical E-Series® Catalyst in the E3 ethylene plant, jointly owned by NOVA Chemicals and The Dow Chemical Company. The E3 Unit operates Front-End acetylene hydrogenation reactors and has set an industry standard in reactor performance by using the superior E-Series® Catalyst technology, combined with a commitment to excellent operation. The catalyst has achieved an unprecedented 18 years of continuous run length without regeneration, while maintaining a consistently high ethylene selectivity. This has improved plant operability, reliability, safety and profitability.

543908 Green Oil Removal from Ethylene Plant Backend Hydrogenation Unit

Abstract:

Proprietary Gas/Liquid Coalescer With Surface Treated Media Cartridges developed, tested, and installed downstream of an Acetylene Hydrogenation Reactor is found to match the removal performance of the Green Oil Wash Absorber; thus, eliminates the need to use recycle wash ethylene solvent stream.

Replacing the Green Oil Wash Absorber with a two-stage Proprietary Gas/Liquid Coalescer System results in increasing ethylene plant production by 5-10% and corresponding increase in plant profitability.

549106 Gas Hydrodynamics of a Novel High-Speed MTO Loop Reactor: The Bypassing and Backmixing Along with Residence Time Distribution

Methanol to olefins (MTO) was an attractive technique to solve such problems as limited oil resource and increasing olefins demands. Many institutions and companies have developed different MTO reactors for industry application. Most of them chose turbulent fluidized bed as the reactors. However there exists a large extent of gas backmixing and long residence time in such reactors, which is detrimental to the MTO reaction. A novel MTO airlift loop reactor with high speed operation condition was developed, with turbulent or fast fluidized bed upward flow in the draft tube in parallel with bubbling or turbulent bed downward flow in the annulus. It aims at improving the flow characteristics of gas-solid phase and optimizing the olefins yield and selectivity. The conversion is a function of the gas hydrodynamics, which is vital for the present research. The novel high-speed MTO loop reactor comprises a fluidized bed and a draft tube for internal circulation of solids, forming separate aeration of the annulus and providing more flexible operation. The gas bypassing between the draft tube region and the annulus region was investigated, aiming at understanding the gas distribution law and qualifying the gas flowrate within the two regions. The gas backmixing in the draft tube region of loop reactor and the average gas residence time in the whole loop reactor were measured by gas-tracing method. The gas backmixing and the average gas residence time in the free fluidized bed were also investigated with the same method to provide a comparison with the novel high-speed MTO loop reactor. The entrained gas from the separation region to the annulus region was analyzed to explain the results of the average gas residence time varying with different operation conditions. Experiment results show that the high-speed MTO loop reactor exhibits little gas backmixing in the draft tube region and shorter average gas residence time compared to that in the free fluidized bed, implying a better conversion and selectivity. Based on the results of those gas hydrodynamics data, the optimal operation conditions can be chosen for the novel MTO reactor.

551144 Experimental and Theoretical Investigation of Radial Slot Jet Flow into Stagnant and Cross Flow Environments

A flowing jet injected into a cross-flowing air stream has several applications in combustion equipment, drying systems, quenching systems, and mixing tanks. A computational fluid dynamics (CFD) technique for simulating radial slot jet flow into cross-flow has been studied and validated using experimental results. Data was obtained from an experimental setup designed and built in the Bio-Energy Systems Technology (BEST) laboratory at Missouri University of Science and Technology. A multi-dimensional hot wire anemometer was used to obtain measurements of radial velocity together with temperature profiles at several axial locations along with center-line axial velocity profiles to study the effect of jet-flow impingement on cross flow velocity profiles. Also, turbulence parameters, kinetic energy and dissipation rate were measured for at different locations. CFD predicted velocity profiles of the system were compared to the experimental data. Measured center-line axial velocity profiles increased significantly immediately after the radial slot injection point. This observation confirmed our ability to control the axial velocity profile by adjusting the cross section jet injection. The velocity data was used to develop a functional relationship between the axial flow and characteristics of the cross flow jet injection design. Good agreement was also found between the predicted and measured flow profiles providing confidence in the CFD model which allows its use for subsequent scaling studies.

543943 Optimizing Polymer Flooding in a Major Heavy Oil Carbonate Field in the Middle East

There are many heavy oil fields that hold vast economic value worldwide. However, the optimal management of these resources remains the main challenge in their development. Therefore, operating companies tend to conduct extensive studies to evaluate the most economic and feasible technical techniques prior to any operational exercise.

This research is concerned with the development of a major heavy oil field in the Middle East. The main objective of the study is to evaluate the applicability of polymer injection in carbonate reservoirs. Experimental work had been conducted as the main methodology utilized to achieve this project target.

Core flooding runs were adopted as the optimal procedure to conduct the experimental evaluation. Both core plugs and oil samples were provided by the operator, while the polymer solution was internally developed by Kuwait Institute for Scientific Research (*KISR*). A total of five different polymer concentration solutions were injected into the plugs to evaluate the behavior of the recovery process.

The selection of the optimal polymer concentration was then achieved by conducting a modeling exercise of the experimental outputs using (*Eclipse-100*) software. The model matching was achieved based on the recovery factor and pore volumes injected for three polymer solutions while the rest of the results were utilized for the model validation.

Several prediction scenarios were run to select the optimal concentration based on the breakthrough recovery factor rather than the maximum one. The reason behind this is that project's economic value normally deteriorates after the breakthrough of the injected fluid into production wells. The study finally concluded that polymer injection is applicable for the studied reservoir and the injected solution of 0.8 % polymer concentration showed to be the optimal.

544842 Desalting Applications

Oil refineries use desalting, liquid-liquid electrostatic precipitation, to remove salt-containing water from crude oil streams. The same technology and equipment is used in oilfield production[1]. In refineries desalting reduces corrosion costs, increases capacity, improves reliability, and makes operation more stable.

This field guide reviews the state-of-the art on application of desalting technology. Topics covered include:

- Desalter types (high-velocity, three-grid, spherical, horizontal)
- Grid types (horizontal, vertical)
- Feed injection (between the grids, below the grids, in the water)
- Feed velocity (adjustable, fixed)
- Mud wash

[1] Oilfield production operations commonly call the operation dehydration instead of desalting.

Garrett Bergquist
Process and Applications Engineer

BHS-Sonthofen Inc.
May 2018

Relationship, Selection, and Optimization of Filter Aid, Filter Media and Clarification Technologies for Contaminant Fines Removal from Process Slurries and Liquids

INTRODUCTION

In many chemical processes there is the need to remove fine solid contaminants less than 5 microrns (μm), from process liquids. When the suspended solids content is dilute, typically less than 1% by weight (wt. %), the removal process is called clarification or polishing filtration to produce a clean liquid product. Typical applications include catalyst recovery, polymer and resins, active pharmaceutical ingredients, removal of unreacted chemicals and byproducts, polishing of grey water, etc.

Often in polishing applications the solids are very fine or amorphous which can be difficult to filter. When filtered, the solids will create a thin coating over the filter media and immediately reduce the filtration rate to an unacceptable level.

In these difficult filtering cases, pretreatment can be used to improve filtration properties and efficiently remove the fine solids from the process liquids. Chemical or physical pretreatment, such as use of flocculants, is one option that can be used to effectively change the nature of the suspended solids. Alternatively, a better filtering medium can be added to the suspension to increase the permeability of the cake. This medium is known as filter aid. An efficient and economical filter aid must have the following properties (World Minerals, 1999):

1. Have rigid, complex shaped, discrete particles
2. Form a permeable, stable, incompressible filter cake
3. Remove fine solids at high flow rates
4. Be chemically inert and insoluble in the process liquid

This paper discusses the use and selection of filter aid and filter media in conjunction with clarification technologies for the removal of fines from chemical process liquids. This article continues with case studies performed to determine the ideal ratios of filter aid for body feed to the suspended solids content as well as to the particle size distribution of these fines.

BHS-Sonthofen Inc.
14300 South Lakes Drive
Charlotte, North Carolina 28273-6794

Tel: 704.845.1190; Fax: 704.845.1902
Website: www.bhs-filtration.com

TYPES OF FILTER AID

Diatomite, perlite, and cellulose are the most frequently used filter aid materials in filtration. Other organic material such as potato starch particles and rice hull ash are also used in the industry but are less common (Sulpizio, 1999).

Diatomite

Diatomite, also known as diatomaceous earth (DE), is comprised of the microscopic silica skeletons of prehistoric plants. These skeletons are called diatoms. This material is mined and calcined to destroy any organic matter. The diatomaceous earth is then milled and separated into various filter aid grades through the use of air flotation (McKetta & Cunningham, 1976). The final product is a soft, powdery material resembling chalk.

Diatomite filter aids are advantageous due to their rigid and highly porous structure. When deposited on a filter media the diatomaceous earth forms a non-compressible cake and acts as a sieve to remove fine solids. The micropores in the diatoms allow flow through the particle while the fine solids are captured between the diatom granules. Being comprised of mostly silica, diatomaceous earth is inert in a wide range of applications.

MARINE DIATOMITE

BHS-Sonthofen Inc.
14300 South Lakes Drive
Charlotte, North Carolina 28273-6794

Tel: 704.845.1190; Fax: 704.845.1902
Website: www.bhs-filtration.com

Perlite

Perlite is formed from glasslike volcanic rock that consists of small particles with cracks that retain water and gas. The perlite is mined and transformed into filter aid by heating the ore to the melting temperature. Once the ore reaches the melting point, the molten volcanic glass expands and fractures due to the emission of the steam and gas. This expanded material is then crushed and classified to provide different filter aid grades.

Compared to diatomaceous earth, perlite has a lower specific weight which allows less filter aid to be used. The primary advantage of using perlite compared to diatomaceous earth is the relative purity. Diatomite may contaminate process liquids with dissolved salts and colloidal clays that are inherent to the diatomaceous earth (Cheremisinoff, 1998). Perlite does not share this risk.

EXPANDED MILLED PERLITE

Cellulose

Cellulose is used as filter aid to a lesser degree than diatomaceous earth and perlite. This is due to its higher cost and lower filtration efficiency. There are some advantages to the use of cellulosic filter aid versus diatomite or perlite. Cellulose is combustible and is useful in the recovery of valuble metals. Cellulose is also compatible with hot caustic solutions where diatomacous earth and perlite are not (Cheremisinoff, 1998).

S.E.M. Images of Diatomite, Perlite, and Cellulose Filter Aids (Sulpizio, 1999)

GRADES OF FILTER AID

For each type of filter aid there are different grades of material which vary in particle size. Typical filter aid particle size ranges from 5μm to 100μm. As filter aid particle size decreases, the ability to capture fine solids increases, thus the filtrate clarity improves. Also as the filter aid particle size decreases, the cake resistance increases. Therefore the finer the filter aid, the throughput per unit area decreases. Given this relationship it is necessary to balance filtrate clarity with filtration rate. It is important to use the largest filter aid grade possible while still achieving the filtrate clarity target.

Since the solids of every application are different, there is no rule for matching filter aid grade to suspended particle size. Instead it is recommended to run benchtop scale filtration tests with various filter aid grades. The rate of filtration can be measured and filtrate clarity observed which will help to select the optimal filter aid product.

FILTER MEDIA

Filter media is the septum on which the filter aid solids are retained while the clear liquid passes through. Most often the media is constructed of a woven synthetic polymer or alloys. The filter media is supported by a rigid supporting structure which is also porous to allow liquid to travel through.

In precoat and body feed applications the filter media pore size should be approximately equal to the median particle size of the filter aid. The exact filter media pore size is not critical so long as some of the filter aid is retained. Once the filter aid begins to build on the woven filter media, the cake becomes the new primary surface for particle capture. Selecting a filter media with too large of a pore size will result in a long turbid cycle as the bridging of the filter aid will take a long time and may be unstable. Selecting a media that has too small of a pore size will result in a reduced filtration flux rate due to the added resistance.

FILTER AID UTILIZATION METHODS

Precoat

In the precoat method, filter aid is used to generate a thin layer of solids on top of the filter media. Once formed, the filter aid cake functions as the primary filter media. Therefore, the filter cloth is no longer the real filter when precoat is used. For that reason, the filter cloth should be as open as possible while still retaining the filter aid material.

The precoat process is achieved by mixing the filter aid into clear liquid or mother filtrate in a precoat tank. A precoat slurry concentration in the 0.3 – 0.6wt% has been found to work best for diatomite filter aids (McKetta & Cunningham, 1976). This slurry is then recirculated through the filter where the solids are captured by the filter media. The clean filtrate is recirculated back into the precoat tank (Figure 1). The precoat thickness should be thick enough to ensure that the entire media surface is coated but thin enough so that it does not provide significant resistance to filtration. In order to prevent the precoat from settling in the pressure vessel and creating an uneven precoat, a feed flow rate of 1 GPM/ft^2 is ideal. Typical precoat thicknesses are in the 1-3mm range for polishing filtration (Svarovsky, 1990).

There are several advantages to using precoat in a clarification processes. Use of a precoat helps prevent the filter media from becoming clogged by the slurry fines, thus extending filter media life. This is especially critical when filtering amorphous or slimy solids. Precoating helps provide immediate filtrate clarity. The precoat layer acts as a depth filtration bed which will help trap fine particles that would typically pass through the filter media. Precoat helps filter cake solids discharge from the filter media at the end of a filtration cycle by providing a cleavage plane from the filter media (World Minerals, 1999).

BHS SONTHOFEN

Body Feed

In body feed filtration applications, the filter aid is blended with the slurry feed either by dosing concentrated filter aid suspension into the slurry feed and in-line mixing (Figure 1), or by mixing the filter aid into the entire slurry batch and maintaining agitation.

The filter aid solids and the native solids continuously build a filter cake until the maximum cake height is reached or the maximum operating pressure is reached. By adding filter aid into the slurry feed, the resulting filter cake is more porous, allowing higher and longer sustained flow rates. Body feed also helps to restrict solids movement which improves filtrate clarity.

The goal of the optimal body feed concentration is to maximize the flow rate of a filtration cycle while avoiding feeding too much filter aid solids. Using too much body feed will cause the filter to reach the maximum filter cake height faster than necessary. When using a body feed without precoat, there can be an initial turbid filtrate while the cake begins to build on the filter media. This initial turbid bypass will generally be much shorter than what would be observed filtering the raw slurry. It is possible to recycle this turbid filtrate back to the slurry feed tank to verify no solids bypass occurs.

Filter Aid Application Summary

	Typical Filter Aid Applications
Filter Media Only	• Crystalline, easy to filter solids • Low solids concentration • Particle size can be matched to cloth pore size • Solids or filtrate is the product
Media & Precoat	• No slimy or amorphous solids • Retention of fines is critical • Solid particles are fine enough to penetrate the precoat bed without blinding precoat OR • Solid particles are coarse enough to deposit on top of precoat without blinding • Filtrate is the product
Media & Body Feed	• Solids are not slimy • Immediate retention of fines is not critical • Solids are difficult to filter
Media & Precoat & Body Feed	• Solids are slimy or amorphous and difficult to filter • Immediate fines retention

UTILIZATION OF FILTER AID FOR CLARIFICATION APPLICATIONS

Filter aid is utilized in clarification filtration processes using two methods: precoat and/or body feed addition. These two methods can be used individually or they can be combined to gain the benefits of both methods. Various filters can utilize filter aid in their filtration cycles. Candle Filter and Pressure Plate Filter are two filtration technologies and are described below. Other filters that can utilize filter aid are filter presses, pressure leaf filters and Nutsche filters.

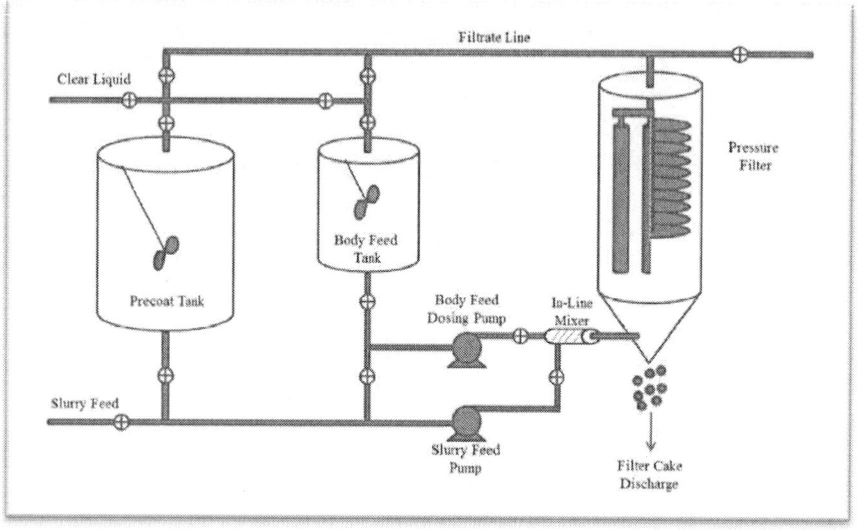

Figure 1: Process flow diagram for precoat and body feed processes with filter aid and filtration via candle or plate filter (Sentmanat, 2011)

Candle Filter Technology

The candle filter is a batch-operated, pressure driven filter. The pressure vessel is filled with tubular filter elements called filter candles. A filter candle is comprised of a dip pipe for the filtrate, a perforated core with supporting tie rods and a filter sock. Each of these candles is installed on horizontal manifolds called registers. Depending on filter size, each register may contain 1 – 20 candles. Given the staggered arrangement and size of the candles on the register, a large filtration area can be provided in a small vessel (Perlmutter, 2015). This arrangement is particularly advantageous in low solids applications <1wt% both with and without filter aid. The large filtration area leads to high throughputs while the cake slowly builds to the maximum filter cake height.

During operation a feed pump forces slurry through a flange into the bottom of the pressure vessel. The solids build up on the outside of the filter sock while the filtrate flows into the candle, through the dip tube and registers. The process continues until the maximum pressure drop, cake thickness, minimum flow rate, or maximum filtration time is reached.

BHS-Sonthofen Inc.
14300 South Lakes Drive
Charlotte, North Carolina 28273-6794

Tel: 704.845.1190; Fax: 704.845.1902
Website: www.bhs-filtration.com

The cake is dewatered by draining the vessel and pressurizing with compressed gas. During cake discharge, the filter socks are inflated by low pressure gas. The process breaks apart the dry cake which falls from the candles and through the discharge valve at the bottom of the vessel.

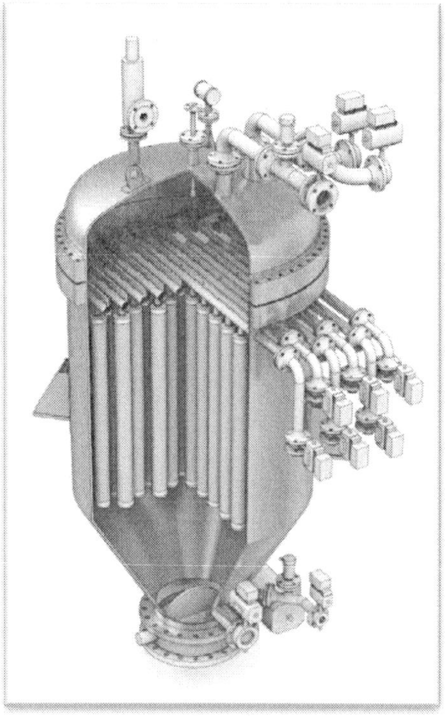

Figure 2: BHS Candle Filter Technology (BHS-Sonthofen, 2017)

Pressure Plate Filter Technology

Similar to the candle filter, pressure plate filters are comprised of filter plates contained within a pressure vessel. Operation is essentially the same as the candle filter with a pump driving slurry into the filter. Solids are captured on the filter media surface while the filtrate is driven away from the vessel.

The main difference between the two technologies is the orientation of the filtration surface and cake discharge mechanism. Instead of vertically oriented candles, the plate filter contains slightly sloped, horizontally oriented, conical-shaped filter plates. The plates have filter media sealed on the top surface of each plate. An opening in the center of the plates allows the filtrate to travel between the plates and out of the vessel. Cake discharge is done by a back pulse of gas in combination with vibration of unbalanced motors. Other cake discharge techniques include spinning plates or manual discharge where the plate stack is removed from the vessel.

Due to the horizontal orientation of the plates, the pressure plate filter is well suited for processes that require a high degree of versatility and cake stability such as incorporating a precoat or cake wash stage. With the vertical orientation of the candle filter, disruption in the positive pressure of the vessel can result in the cake falling prematurely from the media due to gravity. With the horizontal orientation of the filter plates, the cake would be contained on top of the plates if any pressure disruption were to occur. Just like the candle filter, through the use of BHS designed piping and valves the process can be fully automated which mitigates this risk.

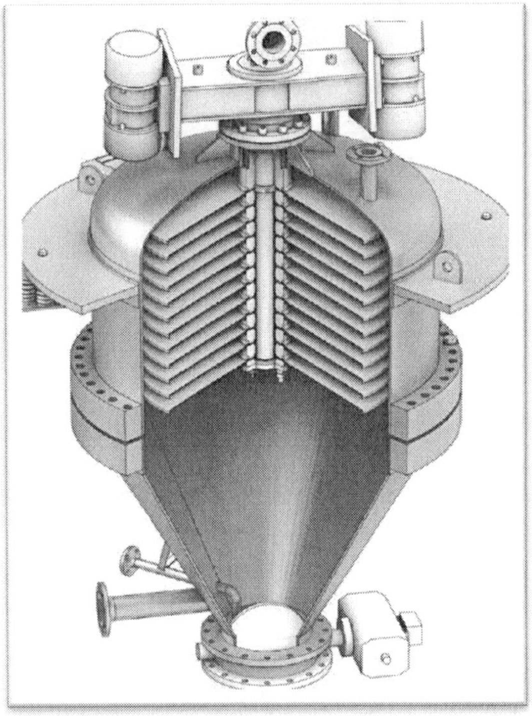

Figure 3: BHS Pressure Plate Filter Technology (BHS-Sonthofen, 2017)

BHS-Sonthofen Inc.
14300 South Lakes Drive
Charlotte, North Carolina 28273-6794

Tel: 704.845.1190; Fax: 704.845.1902
Website: www.bhs-filtration.com

BHS SONTHOFEN

BENCHTOP SCALE STUDY:
OPTIMIZING BODY FEED CONCENTRATIONS IN RELATION TO SUSPENDED SOLIDS CONTENT IN UNTREATED SLURRY

The following study discusses and illustrates a method to physically test and determine the optimal filter aid concentration for a given filter aid grade, filter media, solids size and concentration.

The goal of the case study performed by BHS is to determine the optimal starting point for body feed concentration for suspensions. The tests incorporated slurries comprised of corn starch in water and bentonite clay in water.

Test Setup and Apparatus

The test setup was arranged as shown in Figure 4. A BHS Pocket Leaf Filter (PLF) containing a 12 micron filter media was connected by braided tubing to an agitated slurry tank capable of holding pressure. Hyflo grade diatomaceous earth was utilized as the body feed filter aid for all tests (Table 1).

Median Particle Size	30.1μm
Median Pore Size	7.0 μm
Permeability	1.10 D'Arcys
150 Mesh (% Retained)	7.0 %

Table 1: Hyflo Super-Cel DE Filter Aid (World Minerals, 1999)

The BHS Pocket Leaf Filter apparatus is used for benchtop scale pressure and vacuum filtration tests, to determine flux rates, filtrate clarity, filter media, cake thickness, washing and drying efficiencies, cycle times, performance versus quality parameters and qualitative cake discharge. The PLF is jacketed for heating or cooling and is rated 90 psig to full vacuum. The BHS PLF has a 20cm^2 of active filter area and a filling capacity of 400mL.

The filter medium used for the test series was a multifilament PEEK cloth with a 12μm pore size produced by SEFAR; part number 17-2005-SK012. This filter media was selected due to its high air permeability (150 L/m^{2}s) and small pore size. Due to the high air permeability the resistance from the filter medium is minimized allowing better analysis of the cake resistance. Additionally with a 12μm pore size and an average filter aid particle size of 30μm, the turbid bypass at the start of the filtration cycle is minimized which helps the repeatability of the experiment. In a full scale application, a filter media with a pore size of up to twice the average filter aid particle size could be an appropriate selection.

BHS-Sonthofen Inc.
14300 South Lakes Drive
Charlotte, North Carolina 28273-6794

Tel: 704.845.1190; Fax: 704.845.1902
Website: www.bhs-filtration.com

Batches of slurry were produced by adding measured amounts of corn starch or bentonite into a beaker and mixing with water. Filter aid was added at various concentrations by weight to the suspension and allowed to mix for fifteen minutes. The entire suspension was added to the agitated slurry feed tank and used to pre-fill the BHS pocket leaf filter.

The agitated pressure vessel was used to continuously mix suspended solids and filter aid during the test runs. This vessel was connected to the BHS pocket leaf filter via braided tubing. The pocket leaf filter contained a 12µm woven filter media. Filtration was driven by using 30psig compressed air to pressurize the slurry feed tank which then fed into the pocket leaf filter. After applying pressure to begin filtration, the filtrate mass was measured and recorded in respect to time.

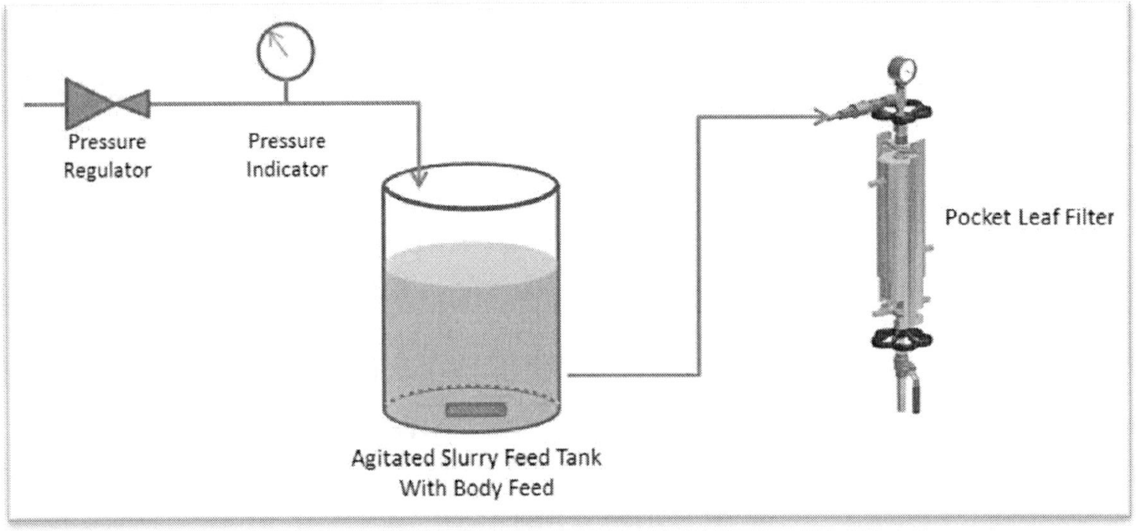

Figure 4: Filtration test setup diagram

Test Results

Corn starch was mixed with ambient temperature water at a concentration of 1wt% suspended solids. Hyflo diatomaceous earth (DE) filter aid was mixed into the suspension at various concentrations ranging from 0wt% to 2wt%. The suspension was filtered at 30psig and filtrate mass recorded versus time. In general, as the body feed concentration increased, so did the filtration flux rate (Figure 5).

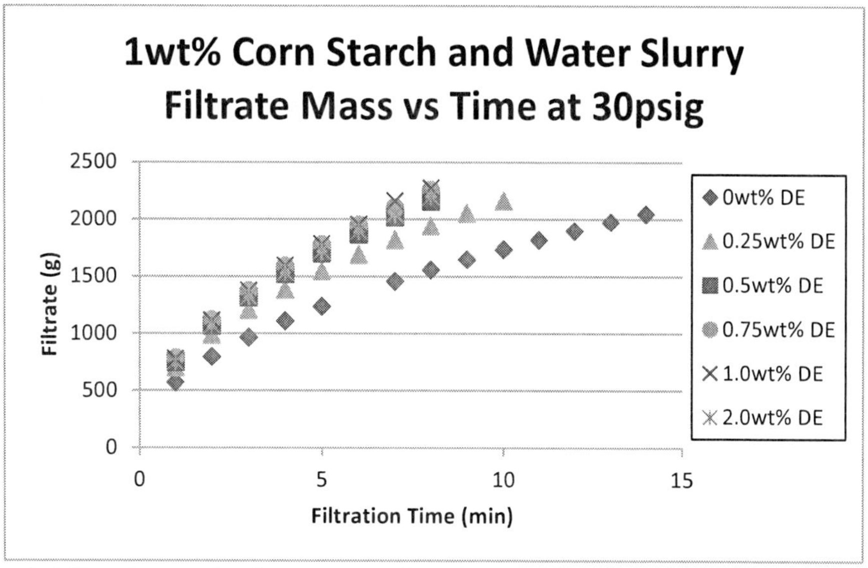

Figure 5: Filtration results for corn starch and water filtration with Hyflo DE body feed at various concentrations

For each test run and data point, filtration flux rate was calculated using the following formula (Equation 1):

$$X(t_{sec})\frac{kg_{filtrate}}{m^2 * s^{0.5}} = \frac{m_{filtrate,t}}{A_{PLF} * \sqrt{t_{sec}}} \qquad \text{(Equation 1)}$$

$X(t)$ = Filtration flux at time (t_{sec})
$m_{filtrate}$ = Mass of filtrate at time (t_{sec})
A_{PLF} = Filter area of PLF test unit ($0.002m^2$)

A plot of the average flux rates for each test at varying body feed concentrations, indicates that the optimal body feed concentration is equal to the original suspended solids content in the slurry.

- In Figure 6, the suspended solids concentration is 1wt%. The highest filtration flux is achieved when the body feed concentration is also 1wt%.
- In Figure 7, the corn starch concentration content in the slurry is 0.5wt%. The highest filtration flux is achieved when the body feed concentration is also 0.5 wt%.

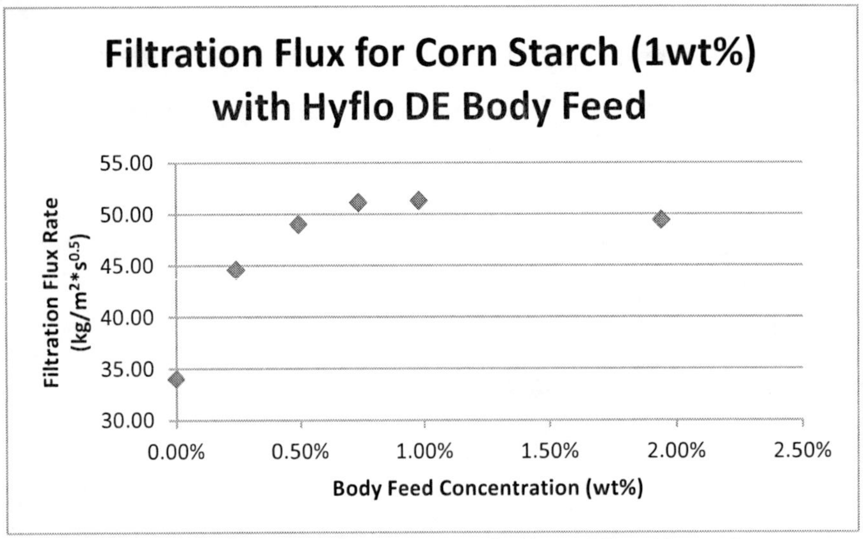

Figure 6: Average filtration flux rate for corn starch (1 wt.%) with various concentrations of Hyflo DE at 30 psig filtration pressure

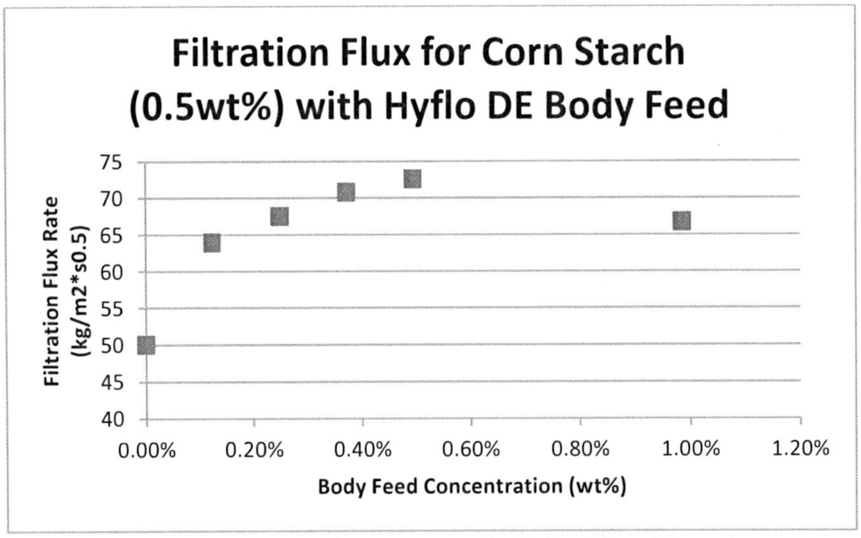

Figure 7: Average filtration flux rate for corn starch (0.5wt%) with various concentrations of Hyflo DE at 30psig filtration pressure

This relationship was also observed and confirmed during the bentonite clay testing. At suspended solids content of 0.25wt% bentonite clay, the maximum filtration flux rate was measured when the body feed concentration was equal to 0.25 wt.% (Figure 8).

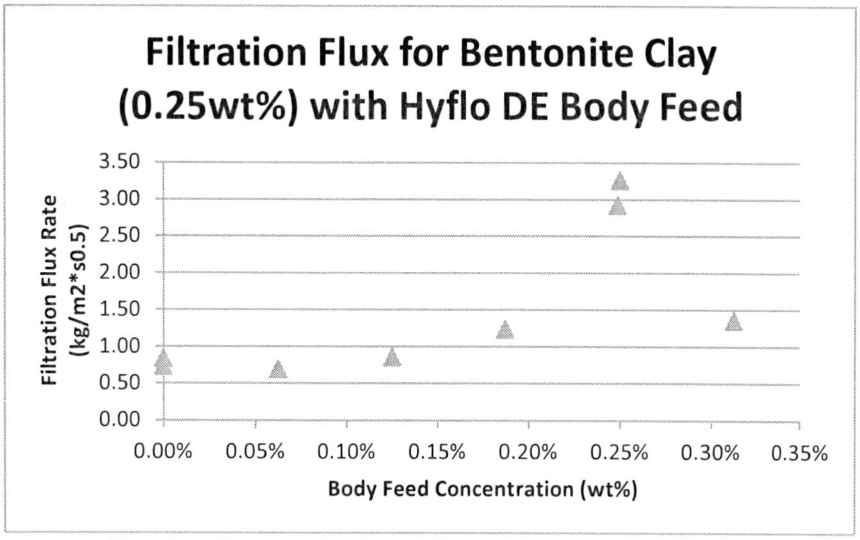

Figure 8: Average filtration flux rate for bentonite clay (0.25 wt.%) with various concentrations of Hyflo DE at 30psig filtration pressure

CONCLUSIONS

Based on the case study results, the optimal starting point of the body feed to suspended solids ratio is recommended to be 1:1 (g filter aid : g suspended solids). This optimal mixture was proven to be true in both the corn starch slurry and the bentonite clay slurries at various suspended solids concentrations.

While this equal parts concentration is optimal for the filtration rate, feeding less filter aid may be more beneficial for reducing filter aid consumption and filtration operating expenses. In the case of the 1wt% corn starch testing, having a ratio of 1:2 (g of body feed: g of suspended solids) resulted in a filtration rate reduction of only 4%. For the 0.5wt% solids test, the same body feed ratio resulted in only a 7% reduction in the filtration rate. By reducing the body feed ratio, there is only a slight compromise to filtration throughput, but a 50% reduction in body feed consumption. Evaluation of the impacts of reducing body feed should be evaluated on a case by case basis to determine the operating cost reduction versus the capital expenditure gains for a larger filter operating at a filtration rate below the optimal point. Reducing the body feed ratio will also decrease the solids load to the filter. In turn this can potentially increase the total filtration time during the filtration cycle; it will take longer to reach the maximum filter cake height. A longer filtration cycle can help improve the run time utilization of the candle filter or plate filter.

BHS-Sonthofen Inc.
14300 South Lakes Drive
Charlotte, North Carolina 28273-6794

Tel: 704.845.1190; Fax: 704.845.1902
Website: www.bhs-filtration.com

New and existing processes looking to introduce body feed into the filtration process should perform similar testing using the BHS Pocket Leaf Filter on a benchtop scale. Performing filtration tests such as these requires little time but can provide significant capital and/or operating cost savings. While this study was limited mainly to filter aid concentration there are many other variables that can be evaluated to optimize the filtration process: filter media, filtration pressure, slurry temperature, filter aid type and grade, use of precoat, precoat thickness, body feed concentration, filtration time/final cake thickness, and many more. All of these variables can be systematically tested using a well thought out test plan.

Results from these benchtop scale tests can easily be translated to a full scale BHS candle filter or BHS pressure plate filter to meet any clarification process needs. These automated technologies provide a significant advantage compared to other pressure filter technologies in the market. Through the use of automation, the BHS filters are able to be used efficiently with no wasted time which leads to a maximum throughput per unit area. With each filtration system built to order, the filters and packages are fully customizable to each process needs.

BHS-Sonthofen Inc.
14300 South Lakes Drive
Charlotte, North Carolina 28273-6794

Tel: 704.845.1190; Fax: 704.845.1902
Website: www.bhs-filtration.com

REFERENCES CITED

BHS-Sonthofen. (2017). *Candle Filter* [Brochure]. Sonthofen, Germany.

BHS-Sonthofen. (2017). *Pressure Plate Filter* [Brochure]. Sonthofen, Germany.

Cheremisinoff, N. (1998). *Liquid filtration*. Boston: Butterworth-Heinemann.

McKetta, J. & Cunningham, W. (1976). *Encyclopedia of chemical processing and design*. New York: M. Dekker.

Perlmutter, B. A. (2015). *Solid-liquid filtration: Practical Guides in Chemical Engineering*. Amsterdam: Elsevier/Butterworth-Heinemann.

Sentmanat, J. M. (2011, October). Clarifying Liquid Filtration: A practical guide to liquid filtration. *Chemical Engineering*, 38-47.

Sulpizio, T. E. (1999, April 6). *ADVANCES IN FILTER AID AND PRECOAT FILTRATION TECHNOLOGY* . In *Advanced Minerals*. Retrieved from: http://advancedminerals.com/pdf/AMC07_ADV._IN%20FILTER_Aid_PRECOAT_TECH.pdf

Svarovsky, L. (1990). *Solid-Liquid Separation*. Burlington: Elsevier Science.

Wakeman, R. & Tarleton, E. (1999). *Filtration: equipment selection, modelling and process simulation*. Oxford, UK New York: Elsevier Advanced Technology.

World Minerals. (1999). *Celite Brand Diatomite Filter Aids* [Brochure]. San Jose, CA: Author.

BHS-Sonthofen Inc.
14300 South Lakes Drive
Charlotte, North Carolina 28273-6794

Tel: 704.845.1190; Fax: 704.845.1902
Website: www.bhs-filtration.com

548017 Gas Liquid Solid High Pressure Separator Using Internal Hot Hydrogen Addition. Part I Experimental

Gas-solid-liquid adiabatic-separator (HCS) was tested in a pilot plant to determine the optimal dimensions and the best operating conditions. The HCS operate to separate catalyst and alkylate from a slurry stream that leave the new alkylation process. HCS use a hot hydrogen stream to reduce density viscosity, vaporize part of hydrocarbon and preheat the solid. Different geometries were test and underflow and overflow stream characterized using different sensors; radial and axial particle size distribution, temperature and pressure were determined using differential pressure and ECT sensors. The experiments were useful to determine the effect of hydrogen flowrate and temperature, as well as the presence of the riser that work as a vortex finder and help liquid-solid separation. The results show that separation efficiency depend on inlet slurry flow rate, cone length, hydrogen flow rate and temperature, solid content and particle size distributions. The conventional isothermal equation fails to simulate the current adiabatic-operation, to predict gas liquid and solid separation efficiencies and device scale-up. Tangential and axial velocity in the lower cone and in the riser were discussed using information obtained by pressure-differential and two ECT calibrated sensors.

543987 Removal of Volatile Organic Contaminants from Air Via Thermal Separation Technique

Thermal separation technique with vacuum de-sublimation is one of advanced methods with industrial processes such as distillation; crystallization and solvent extraction .In the present study an experimental work with simulation model have been carried out to investigate performance of de-sublimator (cold trap) for removal of benzene from gases mixture of BTEX of vented air of a petroleum refinery in Baghdad. Gas chromatography-Mass spectrometer (GC-MS) was used to determine concentrations of Benzene, Toluene, Ethyl benzene and Xylene (BTEX) .The removal efficiency could affected by several operating variables, which are the important task for design and control of the thermal system. Four process variables selected which are: gas concentration; gas flowrate; bulk pressure and bulk temperature. Steady-state solidification process of BTEX was carried out in (0.05 m ID) cylindrical cold trap was made of stainless steel. The thickness of compact solid layer was measured by sophisticated ultrasonic gauge, while pressure and temperature were monitored by high accuracy transducers. Vacuum de-sublimation technique shows one of competitive and economical method when compared with the conventional operations. The bulk temperature has proven the most effective variable and the pressure has critical effect. The predictive mathematical model has been proven reliable for determine unmeasured parameters of heat and mass transfer for the highly interacted non-linear thermal system. Mass transfer operation was the control step at low temperature. Optimum results show that maximum value of removal efficiency (~99.5%) could obtained with high value of benzene concentration (>0.5) and with low values for; gases flow rate (<50 cc/min), bulk temperature (<-6 c) and operating pressure(<3 mbar). Reasonable agreement has obtained when compared the simulated solutions with the experimental results.

.

551352 Highly Graphitic and Weak-Polar Fe-Carbon with Self-Stacked Nanospongy Structure Derived from Silkworm Excrement for Enhanced Selective Recognition of Flavor Anisole

A unique nano-spongy porous Fe-C derived from self-assembly insect-made biomass source (silkworm excrement, SE) was prepared for selective separation of flavor 4-methylanisole (4-MA) from its binary mixture anisaldehyde (AY). We used $FeCl_2/ZnCl_2$ to catalyze SE to a special porous Fe-C (NSCF) by a facile "one-pot" calcination strategy. NSCF exhibited a unique nano-spongy morphology anchoring Fe_3C/Fe_5C_2 in nano-sheeted carbon, which showed an enriched graphitic surface with high dispersion of weak-polar Fe sites. Selective adsorption experiments and DFT simulations demonstrated that 4-MA possessed high electron density in benzene ring and relatively weak polar alkoxy ($-COCH_3$) group, which intensified π-π interaction and weak-to-weak polar attraction with NSCF. This multipiont adsorption made NSCF exhibit an outstanding adsorption capacity (7.3 mmol/g) and selectivity (43.7) at 298 K. This low-cost and eco-friendly carbon from agro-industrial waste can be envisaged as promising adsorbent for highly selective separation of flavoring molecules from hardly separating mixture systems.

557257 Production of Zero Defect (Zd) Slabs Through the Implementation of Smart Manufacturing Technologies in Steel Continuous Casting

Improve steel slab quality and continuous caster productivity using Smart Manufacturing (SM) methodologies and technologies to address the top two KPIs of the continuous casting process – yield (minimize defects/rejects) and uptime (reduce UPTA).

557262 Data Modeling For Machine Learning And Data-Centric Analytics For Smart Aerospace Additive Manufacturing

The goal of this project is to develop technologies on data modeling, machine learning and data-centric analytics for smart Aerospace additive manufacturing and to implement these innovations using data from working Aerospace manufacturing facilities.

557265 Smart Manufacturing For Chemical Processing: Energy Efficient Operation Of Air Separation Unit

The goal of this project is to develop a number of SM Platform™ ready tools and deploy them on Praxair's commercial Air Separation Unit for improving energy efficiency.

557272 Inferential Modeling For Driving Out Energy Waste

The goal of this project will drive out wasted energy in manufacturing facilities through improved information technology. The primary objective of this project will provide technical staff with technology to sniff out unknown or unexpected energy consumption across the facility. Prove the hypothesis that energy waste can be detected without the use of expensive energy meters.

557275 Smart Connected Workers In Advanced Manufacturing

The goal is to develop design tools and open source reference architectures that will enable engineers to reliably assess and implement data flow requirements for affordable, scalable, accessible, and portable smart manufacturing systems (A.S.A.P.) for smart connected workers across all manufacturing sectors.

557276 Energy Management Systems For Subtractive And Additive Precision Manufacturing

The objective of this effort is to mitigate energy waste in manufacturing facilities, and specifically subtractive precision manufacturing, using model-based systems engineering principles.

557278 Smart Manufacturing Of Cement

Using predictive process models, data analytics, sensors and machine learning, a Smart Manufacturing for cement control system platform will be developed in partnership with ARGOS USA cement in an effort to provide a more energy efficient clinker production process with better quality control.
Problem Statement:

557279 Energy Efficient Material Processing Through Automated Process Monitoring And Controls

Demonstrate energy efficient metal material processing at Arconic facility through advanced sensing, automated process monitoring and model based controls

AUTHOR INDEX

AUTHOR INDEX